Culture and Trust in Technology-Driven Organizations

Industrial Innovation Series

Series Editor

Adedeji B. Badiru

Department of Systems and Engineering Management
Air Force Institute of Technology (AFIT) – Dayton, Ohio

PUBLISHED TITLES

Carbon Footprint Analysis: Concepts, Methods, Implementation, and Case Studies, *Matthew John Franchetti & Defne Apul*

Communication for Continuous Improvement Projects, *Tina Agustiady*

Computational Economic Analysis for Engineering and Industry, *Adedeji B. Badiru & Olufemi A. Omitaomu*

Conveyors: Applications, Selection, and Integration, *Patrick M. McGuire*

Culture and Trust in Technology-Driven Organizations, *Frances Alston*

Global Engineering: Design, Decision Making, and Communication, *Carlos Acosta, V. Jorge Leon, Charles Conrad, & Cesar O. Malave*

Handbook of Emergency Response: A Human Factors and Systems Engineering Approach, *Adedeji B. Badiru & LeeAnn Racz*

Handbook of Industrial Engineering Equations, Formulas, and Calculations, *Adedeji B. Badiru & Olufemi A. Omitaomu*

Handbook of Industrial and Systems Engineering, Second Edition *Adedeji B. Badiru*

Handbook of Military Industrial Engineering, *Adedeji B.Badiru & Marlin U. Thomas*

Industrial Control Systems: Mathematical and Statistical Models and Techniques, *Adedeji B. Badiru, Oye Ibidapo-Obe, & Babatunde J. Ayeni*

Industrial Project Management: Concepts, Tools, and Techniques, *Adedeji B. Badiru, Abidemi Badiru, & Adetokunboh Badiru*

Inventory Management: Non-Classical Views, *Mohamad Y. Jaber*

Kansei Engineering - 2-volume set
- Innovations of Kansei Engineering, *Mitsuo Nagamachi & Anitawati Mohd Lokman*
- Kansei/Affective Engineering, *Mitsuo Nagamachi*

Knowledge Discovery from Sensor Data, *Auroop R. Ganguly, João Gama, Olufemi A. Omitaomu, Mohamed Medhat Gaber, & Ranga Raju Vatsavai*

Learning Curves: Theory, Models, and Applications, *Mohamad Y. Jaber*

Modern Construction: Lean Project Delivery and Integrated Practices, *Lincoln Harding Forbes & Syed M. Ahmed*

Moving from Project Management to Project Leadership: A Practical Guide to Leading Groups, *R. Camper Bull*

Project Management: Systems, Principles, and Applications, *Adedeji B. Badiru*

Project Management for the Oil and Gas Industry: A World System Approach, *Adedeji B. Badiru & Samuel O. Osisanya*

Quality Management in Construction Projects, *Abdul Razzak Rumane*

Quality Tools for Managing Construction Projects, *Abdul Razzak Rumane*

Social Responsibility: Failure Mode Effects and Analysis, *Holly Alison Duckworth & Rosemond Ann Moore*

Statistical Techniques for Project Control, *Adedeji B. Badiru & Tina Agustiady*

Culture and Trust in Technology-Driven Organizations

Frances Alston

CRC Press
Taylor & Francis Group
Boca Raton London New York

CRC Press is an imprint of the
Taylor & Francis Group, an **informa** business

CRC Press
Taylor & Francis Group
6000 Broken Sound Parkway NW, Suite 300
Boca Raton, FL 33487-2742

First issued in paperback 2019

© 2014 by Taylor & Francis Group, LLC
CRC Press is an imprint of Taylor & Francis Group, an Informa business

No claim to original U.S. Government works

ISBN-13: 978-1-4822-0923-5 (hbk)
ISBN-13: 978-0-367-37913-1 (pbk)

Visit the Taylor & Francis Web site at
http://www.taylorandfrancis.com

and the CRC Press Web site at
http://www.crcpress.com

Contents

Preface

Changes in the global business environment have drastically increased demands on businesses in the areas of productivity, product quality, innovation, and product development. Competing in a global environment has increased the need for enhanced communication exchanges, creation of trusting cultures, and the ability to cope with the increased complexity and uncertainty of the ever-changing business environment. It is widely recognized that culture and trust play important roles in the corporate environment and must be managed in order to recruit and retain talented workers and build high-performing organizations. Managers have a need to understand the relationship between culture and trust and the role each plays in the successful creation of high-performing organizations. Authors and theorists have linked trust with increased performance levels, creativity, and critical thinking. These factors are critical for leaders to tap into when work is being performed in flexible and adaptive environments. Many researchers, authors, and theorists have contributed to developing the theories of culture and trust and have pointed out the important role that culture can play in the success of organizations.

Presented in three parts, this book is written for the theorist, the researcher, and the practitioner. Section I outlines the literature on organizational trust and culture and the role theorists believe these play in the success of a changing domestic and global business environment. In conducting the literature review, various elements or attributes of culture and trust were studied. These attributes include the various ways of defining culture and trust and the survey instruments used to measure culture and trust. The results of two studies that demonstrate the connection between organizational culture and trust are addressed in Section II. The two studies were conducted at separate times using data collected from several companies within a three-hour radius of one another. These companies are highly dependent upon the ability to identify, hire, and retain highly skilled knowledge workers. These workers are critical for the companies to compete successfully within the scope of their business and expand into their current and other markets. The result of the two studies sought to identify the relationship between organic/mechanistic cultures as described by Burns and Stalker and trust within organizations. The two empirical studies investigated the relationship between mechanistic and organic cultures and the level of trust in technology-based organizations.

Data were collected from more than 650 survey participants who were employed by technology-based organizations. Each participant completed three survey instruments designed to measure organizational culture and trust, and collect pertinent demographic characteristics. The results of both studies showed that there is a strong correlation between mechanistic and organic cultures and the level of trust present in organizations. The results also showed that the more organic an organization is perceived, the higher the level of trust that is present. The analysis of the attribute of trust studied (openness and honesty, competence, concern for employees, and reliability) showed a strong correlation with the culture of an organization.

Section III of this book was written to assist practitioners in their efforts to develop and improve trust in their organizations through designing an appropriate culture that is supportive of building and sustaining trust. Section III also contains a practitioner's guide, containing tools managers can use to assess, diagnose, and improve employees' perception of their work culture, thereby improving the level of trust found in organizations. This guide serves to provide management with actions and activities that should be considered when handling the day-to-day business of the organization. If followed, these activities can be instrumental in designing a culture that leads to success and ease of operation for the organization and its members.

In addition, the study provided insight into the type of culture that typically can be found in technology-based organizations. The organizations that participated in the study cultures were not completely organic or mechanistic. These organization cultures were aligned at various locations between mechanistic and organic on the culture continuum. This finding provided the means to define another type of cultural environment for organizations exhibiting cultures such as the ones illustrated during the study. This book concludes with a section that can be used to assess an organization and one's own actions and skills.

About the author

Frances Alston, PhD, has built a solid career foundation over 25 years while leading the development of, managing, and implementing environment, safety, health, and quality (ESH&Q) programs in diverse cultural environments. This includes working with diverse teams in developing ESH&Q programs and fostering a safety posture required for working in high hazard environments with specialty chemicals and radioactive materials. She designed and championed implementation of an occupational hygiene program based on the European Health and Safety Standards used in the United Kingdom. Throughout her career, she has delivered superior performance within complex, multi-stakeholder situations and has dealt effectively with challenging safety, operational, programmatic, regulatory, and environmental issues.

Dr. Alston is effective in facilitating integration of ESH&Q programs and policies as a core business function while leading a staff of business, scientific, and technical professionals. She is skilled in providing technical expertise in regulatory and compliance arenas as well as when determining necessary and sufficient program requirements to ensure employee and public safety, including environmental stewardship and sustainability. She also has extensive knowledge and experience in assessing programs and cultures to determine areas for improvement and development of strategy for improvement.

Dr. Alston holds a BS in industrial hygiene and safety, an MS in hazardous and water materials management/environmental engineering, a MSE in systems engineering/engineering management, and a PhD in industrial and systems engineering.

She is a fellow of the American Society for Engineering Management (ASEM) and holds certifications as a Certified Hazardous Materials Manager (CHMM) and a Professional Engineering Manager (PEM). Her research interests include investigating and implementing ways to design work cultures that facilitate trust.

## chapter one

Importance of trusting cultures

1.1 Introduction

Trust is becoming more important to the success of corporate business strategies as companies continue preparing to compete in the changing global economy. Many companies have seen trust erode over the years due in part to the unethical actions of some business leaders. Workplace trust can be broadly defined as having the belief that a person or an employer will be honest and consistently follow through with commitments. Trust is a central element in designing the culture of an organization and is considered to be the foundation of high-performing organizations and teams, because trust can significantly influence business outcomes. Many positive attributes of an organization have been linked to trust, including increased performance levels, increased creativity, and critical thinking, all of which are necessary factors when work is being performed in flexible and challenging environments. Organizational trust is deemed necessary for businesses to survive economically and to compete successfully in cultural environments subjected to the dynamics of a constantly changing marketplace.

Culture provides the context in which the organization and its members function, build relationships, and perform their assigned tasks. Historically, culture has been broadly described by many as a group of patterns, beliefs, behaviors, and values that function as an integral part of organization systems and determine the actions of its members. The success of individuals as they implement new processes is a function of how they relate to the culture of that organization. Cultural features affect group members internally at the values and beliefs level, as well as externally at the working level, therefore, culture can be the single most important factor in determining the level of trust found in organizations.

Many researchers, authors, and theorists have contributed to developing the theories of culture and trust and have alluded to the important role that culture may play in the trust-building process. Cultural values can directly influence the erosion or development of trust; therefore, the culture of an organization can affect people's willingness and ability to trust. As a result, culture can be viewed and used as a tool to unite or divide people, teams, and organizations. Culture can also affect people's perceptions of fairness and equity in the decision-making process. Thus, it is important for managers to understand better how trust is developed and the ways culture can influence the trust-building process. The literature review and empirical research discussed in this book seek to build a better understanding of the relationship between organizational culture and the level of trust found in technology-based organizations. Although the two studies discussed in this book focus on technology-based organizations, the results and knowledge gained through the studies can be applied to businesses globally.

1.2   Technology-driven organizations in the global marketplace

According to the National Science Foundation there are very few world events that are not technology driven or influenced by various aspects of technology. During the years preceding and into the 21st century, the global economy has been in a rapidly evolving state with a rapidly changing structure, and technology is playing a key role in making the changes successful with the development of new concepts, processes, technologies, and products. The global market for high technology continues to grow at a rapid pace and high-technology industries are at the forefront in contributing to economic growth worldwide. High-tech industries are firms that typically compete in fields such as engineering, computers, R&D, information technology, chemicals, pharmaceuticals, and instrumentation. The National Science Foundation lists the following reasons for the importance of high-technology firms in the global economy.

1. High-technology firms tend to assume an innovative posture, and firms that innovate tend to gain and sustain market share, as well as create new product markets, while gaining efficiencies in productivity.
2. High-technology firms engage in the development of value-added products and are generally successful in foreign markets, which typically results in higher compensation for their employees.
3. Research and development (R&D) activities performed by high-technology firms tend to benefit other commercial sectors by generating new products and processes that contribute to the increase of productivity, expand business, and create higher paying jobs.

From 1980 to 1998, high-technology products grew at a rate of nearly 6.0% per year compared to the approximate 2.7% growth rate for other manufactured goods. Businesses across the globe are being forced to make adjustments to accommodate the changing global economy in order to maintain their competitiveness. During the 1980s, the United States and other countries began devoting more resources toward the manufacture of technology-related goods.

Employment in high-tech industries increased approximately 7.5% during 1992 to 2002. High-tech employment realized continual growth worldwide during 2012 while adding jobs to the national economy. Technology-based organizations typically employ a large number of highly skilled workers from the fields of science and engineering. Technology-oriented workers are important assets for technology-based organizations because these professionals are engaged in activities such as R&D, developing product lines, production process changes and upgrades, and new equipment design. Technology-oriented professionals are well educated, generally hold advanced degrees as well as other professional credentials such as certifications, and possess a high degree of intelligence and desire for learning. They tend to be creative and seek out challenges and new ways of resolving complex issues.

Science and technology are widely regarded as important for the growth and competitiveness of individuals and for overall national economic growth. Leaders in developing and developed countries are striving to attract, cultivate, and retain knowledge-based companies and workers and they continue to emphasize the important role of knowledge, particularly R&D activities and other activities requiring advanced science and technology, in a country's economic growth and competitiveness. Knowledge- and technology-intensive (KTI) industries are considered to have a strong link to science and technology. Specifically, these industries are important to developing technological infrastructure that diffuses across the entire economy. The globalization of the world's economy involves the rise of new centers of KTI industries. Although the United States continues to be a leader in KTI industries, developing economies, such as Asia, are actively engaged in the pursuit of innovation. KTI industries have become a major part of the global economy and represent a growing share of many countries' total economic activity. The global value added to these industries totaled about $18.2 trillion in 2010 and global value-added output for high-technology manufacturing industries increased from about 700 billion in 1995 to about 1.4 trillion in 2012.

Productivity growth is essential for advancing and maintaining our standard of living. Therefore, greater productivity is important because it raises workers' income levels and allows the economy to grow without inflation, which can assist in keeping interest rates low. Improvements in productivity growth are essential in improving the status of the nation in terms of developing and sustaining jobs, as well as maintaining and improving domestic and international competitiveness. Productivity gains are important in ensuring long-term vitality of the national economy and in increasing a company's profitability. Increased productivity is one of the best ways to secure improvement in the quality of life. Therefore, for firms to remain competitive and successfully increase their profitability, they must have the ability to improve and sustain high productivity growth. Growth in new technology continues to yield strong productivity and profit growth and due to the globalization of high technology, competition is strong. High-technology industries continue to receive a great deal of attention because of the role these industries play in the economic growth of the nation.

1.3   Relationship between increased productivity and trusting cultures

Literature written throughout the 1990s and 2000s has consistently linked high levels of trust with greater performance. A trusting workplace is viewed as being instrumental to the success of organizations and trust is credited with enhancing process improvements, increasing productivity, and efficiency. Trust is also credited with contributing to increasing positive workplace behaviors, such as positive attitudes, higher levels of cooperation, high levels of performance, better team interaction, and increased employee morale. Organizational trust can serve as the catalyst that influences productivity and job performance by creating an environment that encourages cooperation, which allows employees to concentrate on the task at hand. Thus, developing trust in the workplace is critical to the success of companies.

Many believe that trust can have a significant impact on organizational success. The environment of trust within an organization is greatly affected by the culture of that organization. Business outcomes can be significantly influenced when trust is treated as a cultural characteristic of the organization. Therefore, there is a need to create a culture that develops, reinforces, and sustains trust. It has been stated that high-trust cultures usually have fewer and less rigid controls in place. Conversely, low-trust cultures are characterized by bureaucratic policies and procedures that limit individual creativity and initiative. Burns and Stalker (1961), during their research into organizational systems, identified three organizational environments in which work was performed: stable, changing, and innovative. In stable environments tasks were firmly established and performed generally through written instructions. The hierarchical management systems in these environments were devised to ensure that production and production conditions remained constant. Communication in stable environments was characterized as having limited information flow. Changing environments were characterized as having hierarchical management systems. Design and production remain relatively standard and management generally involves workers in solving problems. In innovative environments specialization of tasks was avoided to allow for task adjustments as needed. Total dependence on management for authority and defining task functions was discouraged. Management listens to and includes workers in the decision-making process. These three environments formed the basis for the two divergent cultures later referred to as mechanistic and organic.

Mechanistic and organic culture descriptions have been used by many to describe both culture and organizational systems and are characteristically aligned with what is believed to exist in high- and low-trust organizational environments. Mechanistic cultures are characterized by having jobs that are rigidly defined and tend to be governed by procedural compliance, with communication flowing from the top down to lower levels in the organization. Organic cultures are characterized as having jobs that are flexible and adjusted as needed with fluid information flow across the organization. Organizations with organic cultures are also characterized by risk-taking and valuing employees. High-tech firms that are competitive tend to have cultures that are flexible, adaptable, foster teamwork, and promote risk-taking.

Depending on the cultural values of an organization and the actions of the leadership team, trust initiatives may be helped or hindered. If technology-based organizations are to thrive in the global economy, trust must be an integral part of the culture of the organization. In high-trust environments, people are more willing to share information, admit to and learn from mistakes, and take on challenging tasks. In work environments where trust flourishes, the stage is set for improved morale and productivity. It is widely recognized that a culture of trust can have a significant impact on an organization and culture can directly affect an individual's willingness to trust.

A better understanding of the relationship between organizational culture and trust in technology-based organizations is important because increased productivity has been linked to a high-trust cultural environment. In order to thrive in the global economy of the 21st century, high-tech organizations must be able to adapt to the rapidly changing conditions of the marketplace. Organizations with cultures that are flexible require a trusting workforce to succeed.

section one

Literature review and the empirical study

Section I outlines the literature review of the theoretical information found in the area of organization culture and trust. An in-depth review of the various ways the two concepts are defined, the prevailing knowledge, and a discussion of the types of survey instruments available to measure organizational culture and trust are also discussed in detail.

chapter two

Literature review

2.1 Introduction

An extensive literature review was conducted to identify existing theoretical and empirical knowledge in the areas of organizational culture and organizational trust. The literature shows that culture and trust can both play distinctive roles in an organization. Organizational culture has a powerful effect on the performance as well as the long-term effectiveness of an organization. The reviewed literature contains a plethora of information on the culture of organizations. The culture literature has focused primarily on the basic values, beliefs, behavior, and practices that are inherently present in organizations.

The literature describing organizational trust has made some advances in contributing to the understanding of how trust relates to certain facets of an organization. Trust is realized and strengthened by social interaction, cultural affinity between people, and the support of group norms and behaviors. Organizational trust is viewed as a multi-level phenomenon that is closely related to norms, values, and belief in the organizational culture. An organization's culture based on trust fosters attachment and facilitates relationship building among individuals. A trusting culture facilitates learning and innovative culture where people are allowed to make mistakes and learn from them.

2.2 Organizational culture

Many authors and theorists have contributed to organizational culture research. Edgar Schein (1992) can be noted as one of the most significant contributors. His work is used as the cornerstone for others who continued the evolution of organizational culture theory. Schein formally defines culture as "a pattern of shared basic assumptions that the group learned as it solved its problems of external adaptation and internal integration that has worked well enough to be considered valid and, therefore, to be taught to new members as the correct way to perceive, think, and feel in relation to those problems."

The concept of culture in the organizational and management literature originated from various anthropological and sociological sources. Organizational studies of culture suggest that culture can be created to exhibit the creator's values, priorities, and vision. Pragmatists generally see culture as a key to commitment, productivity, and profitability. They collectively argue that culture can and should be managed. In doing so, collective belief systems about social arrangements are developed. Philosophers tend to call these shared systems shared paradigms, whereas sociologists speak of them as social reality, and anthropologists call them cultures. The most central aspect of culture involves beliefs about the "appropriate" nature of transactions. Whenever a transaction takes place, valued things such as facts, beliefs, perceptions, and ideas are exchanged. These transactions or exchanges determine identity, power, and satisfaction. The governing rules about the nature of transactions tend to reflect deeply embedded values. They include beliefs and perceptions about organizational purposes, the level and appropriateness of authority, the decision-making process, leadership style, and management competence.

Substantial influence can be placed on an organization by its culture, because the shared values and beliefs that are present within a culture represent important variables that guide behaviors. Culture influences the behavior of members within an organization and affects many aspects of organizational life, such as the distribution of rewards, promotion, and how people are treated. Culture, therefore, has become an important element in understanding organizational processes. The literature on organizational culture and performance of organizations is interpreted as advocating culture as a positive economic value for a company. A company's culture can be a viable source of sustainable competitive advantage only if the culture itself is appropriate and valued.

2.3   Defining organizational culture

Organizational culture is defined differently by various authors, theorists, and practitioners. Gudykunst, Stewart, and Ting-Toomey's (1985) review of the cultural literature led to a broad definition of organizational culture as the process of (1) regulating and generating meanings, as a system of shared expectations for behavior; (2) relating shared practices such as ritual, myths, and beliefs to organizational culture; (3) identifying cultural phenomena as communicatively constituted in organizational disclosure; and (4) utilizing and embracing qualitative analyses.

Deal and Kennedy (1982) describe culture in terms of its strength. They state that a strong culture is one with formal rules that list exactly how people should behave. Knowing exactly what is expected, employees will most likely minimize wasted time in determining how to react in a given situation. For example, in a weak culture, time is wasted in trying to ascertain how and when to respond. According to George and

Jones (1996), culture is an informal set of values and norms that controls human interactions within an organization. Bender (2000) describes organizational culture as the shared underlying assumptions and core values, the behaviors and habits (the way business is conducted), the symbols and language (tangible aspects such as company logos, dress attire, songs, etc.) present in organizations.

Some theorists define organizational culture in terms of attributes or elements that are critical to an organization's success. Culture is defined by Harrison and Stokes (1992) as the patterns, values, rituals, myths, and sentiments that members of an organization share. Heskett and Kotter (1992) envision organizational culture as an entity having two levels. These levels are shared values and group behavioral norms. Shared values consist of the goals that the people in the organization share that are instrumental in shaping behaviors. Group behavioral norms are viewed as common ways people are expected to act. These actions or behaviors serve as a teaching mechanism for new members. Schien (1992) proposes three elements of culture that include artifacts, espoused value, and basic underlying assumptions. Alvesson (1993) defines culture in terms of four attributes: (1) the degree to which an organization is unique in developing its organizational patterns, (2) the degree to which the organization is viewed as a cohesive entity, (3) the degree to which it is seen as being independent of external environment influences, and (4) the appropriate level for highlighting the culture phenomena. Trice and Beyer (1993) support a concept that lists four major categories or elements of cultural forms known as symbols, language, narratives, and practices.

Burns and Stalker (1961) define culture in terms of management systems that are referred to as mechanistic and organic cultures. Organizational characteristics determine whether an organization management system is viewed as being mechanistic or organic in nature. The Burns and Stalker system is further defined in Section 2.4. Denison (1990) derived a definition of organizational culture that ties the cultural values of an organization to the management systems operating within that organization. He defines organizational culture as "the set of values, beliefs, and principles that serve as the foundation of an organization's management system along with the practices and behaviors that reinforce them."

A summary of the various culture definitions and attributes are summarized in Table 2.1. The varying ways of defining culture affirmed that many authors are in agreement that culture in the early years referred to the practices, values, beliefs, and behavioral patterns that form the identity of an organization. In later years language and symbols were included as important attributes in defining culture.

Table 2.1 Culture Definitions Attributes

Author	Practices	Values	Beliefs	Behavior	Language	Symbols
Burns and Stalker, 1961	X	X	X	X		
Deal and Kennedy, 1982	X			X		
Sathe, 1983			X			
Robbins, 1984	X					
Gudykunst, Stewart, and Ting-Toomey, 1985	X		X	X		
Amsa, 1986	X	X	X			
Harris, 1989	X			X		
Denison, 1990	X	X	X	X		
Sackmann, 1991		X	X	X		
Harrison and Stokes, 1992	X	X		X		
Hampden-Turner, 1992	X		X			
Johns, 1992		X	X			
Kotter and Heskett, 1992		X		X		
Schein, 1992	X	X	X			
Alvesson, 1993	X					
Pheysey, 1993	X					
Trice and Beyer, 1993	X				X	X
Fairholm, 1994	X					
George and Jones, 1996	X	X				
Pearce and Robinson, 1997	X					
Daley and Vasu, 1998	X	X	X			
Bender, 2000		X		X	X	X
Hofstede, 2001			X	X		X
Schein, 2010	X	X	X	X		X

2.4 Burns and Stalker's organic and mechanistic management systems

As stated above, Burns and Stalker have defined culture in terms of its prevailing or dominant practices and management systems, and refer to these practices and systems as mechanistic and organic cultures. Mechanistic cultures are generally characterized by highly specialized jobs, detailed job descriptions, functional division of work, centralization, procedures, and vertical communication. In mechanistic environments, management strongly believes in and places value in loyalty and obedience to supervision. Work behaviors in mechanistic settings are controlled by instructions and all decisions are issued by supervision. General organizational practices include differentiation of tasks with hierarchical structures of communication and authority. These types of cultural environments are highly adaptable to stable environments and are considered optimal environments for technology implementation. An organization that operates in a mechanistic culture tends to follow the rational model style of management, which endorses the need to analyze and control everything and to adhere closely to the chain of command. According to Tschannen-Moran (2001), organizations tend to use rules and regulations when there is a lack of trust among their members.

Organic cultures are characterized by having open lateral communication, jobs that are not clearly defined and are adjusted as needed, little preoccupation with following the chain of command, and communication across all levels of the organization. In organic environments, management believes in and practices redefining and adjustment of tasks as needed and places high value on employees being committed to the task at hand. Schoderbek and Cosier (1991) believe that organic cultures exhibit additional characteristics such as tolerance for diversity, minimal defensiveness in interpersonal relationships, open confrontation of issues, respect for individuality, and flexibility. Organic structures are viewed as highly adaptable to unstable changing environments and are optimal for technology development.

2.5 Likert System IV management process

The Likert IV Management Process defines organizational practices in terms of the management processes or systems and their relationship to subordinates within organizations. The Likert system is divided into four categories: System I (Exploitive–Authoritative), System II (Benevolent–Authoritative), System III (Consultative), and System IV (Participative). Likert System I organizations generally exhibit characteristics such as support for management's use of threats to motivate workers into accomplishing work and topdown communication, with most of the decisions being made at the top of the organization. The organizational practices noted by the Likert System I are consistent with the characteristics outlined by Burns and Stalker as being prevalent in mechanistic organizational environments.

Likert System II organizations are characterized by a lack of commitment between lower and upper levels of management. Management uses rewards to achieve the appropriate or desired behavior. Communication is mostly generated at the top with some upward flow of information. Decisions are made primarily at the top of the organization with some decisions being delegated at or to lower levels in the organization.

Likert System III organizations operate in an environment where management offers motivation through rewards and occasional punishment for those who defy or do not support authority. Big decisions come from the top down to lower levels of the organization. Management often listens to employees, but reserves the right to make the final decision. There is some reliance by management on intrinsic rewards, however, extrinsic rewards are most often used. Most of the employees generally exhibit favorable attitudes toward management.

Likert System IV organizations are characterized as having highly productive employees. Management encourages participation and involvement in setting organizational goals and performance and intrinsic rewards are commonly used. Decision-making occurs through group processes where each group is linked by individuals who are members of more than one group. Management and workers have a close working relationship utilizing a team-based approach to problem solving and communication flows in all directions across the organization. The organizational practices noted by the Likert System IV are consistent with the characteristics outlined by Burns and Stalker as being prevalent in organic organizations.

2.6 Connection between mechanistic and organic cultures and Likert management systems

Authors and theorists collectively believe that the cultural characteristics of an organization are related to the management systems operating within that organization. Likert developed a management system that can be used as an indicator of subordinate relationships in organizations. The Likert system emphasizes the importance of the organizational structure and the cultural environment. During previous years others have used the Likert Systems I and IV interchangeably with Burns and Stalker's mechanistic and organic cultures. Likert also developed a survey instrument to measure Systems I to IV that is widely recognized and extensively used. Research conducted by Burns and

Stalker demonstrated that the management processes in organizations are related to their culture. Their research explained the relationship between organizations and their operational environments resulting in the two divergent management systems known as mechanistic and organic cultures. Theorists share Likert's view that the culture of an organization is linked to the prevailing processes and systems operating within the organization.

The Likert System I is considered to be analogous to Burns and Stalker's mechanistic culture, displaying the same organizational characteristics typically found in organizations having mechanistic cultural environments. Both organizational environments depend heavily on involvement of upper management in the decision-making process and channeling of communications. In these environments motivation is achieved generally through fear. Close adherence to the chain of command and little or no teamwork among organizational members are prevalent in mechanistic environments.

The Likert System II shares many elements of Burns and Stalker's mechanistic cultural environments albeit at a more moderate level than those found in System I. These mechanistic characteristics are found and implemented in varying degrees. Mechanistic elements present in System II environments include some upward communication, but information flows primarily down to lower levels of the organization from top management, and motivation is accomplished through the fear of being punished. However, these shared characteristics are not practiced at the same level as practiced in Likert System I (mechanistic) environments. For example, in System I environments most of the decisions are made at the top of the organization. In System II environments, for the most part, decisions are made at the top of the organization with a few decisions being made at lower levels of the organization.

The Likert System III exhibits many characteristics of an organic cultural environment and includes the presence of some upward communication, fluid information flowing to upper management as well as to lower levels of the organization, and moderate usage of teams to solve problems. Although System III is closely aligned with organic environments, there are some differences in the level or degree to which the various elements in common are practiced. For example, in System III environments communication within the organization flows vertically as well as horizontally, whereas in System IV (organic) environments communication flows vertically, horizontally, and laterally (with peers) throughout the organization.

The Likert System IV is analogous to an organic environment, with both exhibiting comparable organizational characteristics. Both the Likert System IV and organic cultural environments recognize the importance of all employees supporting the team-based approach to problem solving and communications flowing in all directions. These practices are prevalent in the culture of the organization. In these organizational environments group participation is highly visible. The relationship between Burns and Stalker's mechanistic and organic cultures and the Likert System is summarized in Table 2.2.

The Likert Systems I and IV and Burns and Stalker cultural environments display common organizational characteristics. The Likert System I is analogous to the characteristics outlined in a mechanistic cultural environment, whereas the Likert System IV is analogous with the characteristics found in organic cultural environments. Likert Systems II and III share some characteristics of mechanistic and organic cultures. These systems can be said to reside between mechanistic and organic cultures on

Table 2.2 The Likert Management Style[a] and Organic/Mechanistic Cultures[b]

Management style/culture	Characteristics
Likert System I (Exploitive–Authoritative) *Mechanistic*	Vertical communication (downward). Practically no teamwork. Decisions made at the top of the organization. Work governed by procedures or directions from supervisor. Close adherence to chain of command. Highly specialized task structure.
Likert System II (Benevolent–Authoritative)	Leader uses rewards to encourage appropriate performance. Slight amount of teamwork. Communication flows down from management. Listens somewhat to concerns lower in the organization. There may be some delegation of decisions. Almost all major decisions are still made centrally.
Likert System III (Consultative)	Some upward flow of information and efforts to listen carefully to ideas. Moderate amount of teamwork. Communication flows down and up. Major decisions are still largely made at the top.
Likert System IV (Participative) *Organic*	Leader makes maximum use of participative methods (teamwork). People participate in decision making. Communication flows in all directions. Continuous adjustments of tasks as needed.

Source: (a) Likert, R., *The Human Organization: Its Management and Value*, McGraw-Hill, New York, 1967. (b) Burnes, T., and Stalker, G.M., *The Management of Innovation*, Tavistock, London, 1961.

the culture continuum. The Likert System I was used as representative of the Burns and Stalker mechanistic cultural environments and the Likert System IV was used as representative of an organic cultural environment for the purpose of measuring culture in this study.

2.7 Summary of culture literature review

There are many definitions of organizational culture; however, these definitions generally refer to the shared meaning (behaviors), values, beliefs, and expectations or practices within the organization. Theorists believe that the cultural characteristics of an organization are related to the organization's management system. It is theorized that trust makes organizations more organic, eliminating the need to rely on the abundance of impersonal rules to manage in changing environments and that high-trust cultures usually have fewer and less rigid controls in place. Organic cultural environments as defined by Burns and Stalker can be defined in terms of Likert's System IV management process, whereas mechanistic cultures can be defined in terms of the Likert System I management process. Culture was assessed for the purpose of this study as being on a continuum from mechanistic to organic and was measured using Likert Systems I and IV.

2.8 Organizational trust

Trust is globally viewed as a social expectation that has to do with people's perception of the integrity/honesty, caring, and competence of an individual or system that is verified by experience. Trust is a condition of situations and of human relationships, therefore it is surmised that organizational situations can encourage or discourage trust. Trust is essential in developing mutually dependent relationships and is based upon repetitive actions that yield constant results. The most frequent triggers of mistrust, according to Ryan and Oestreich (1998) which were identified through extensive field investigations were as follows: (1) management and supervisor displays of abrasive and abusive conduct, (2) ambiguous behavior by management and supervisors, and (3) the employee's perceptions about the organization's culture. The organization's human resource system functions and the behavior of leaders generally will provide clues about the nature of the culture and whether the culture is fear-based or trust-based. The perception and actions of the leadership team can trigger responses of fear and mistrust.

It is easier to achieve trust when an organization's aim, vision, mission, values, objectives, and goals are understood and shared. Being honest and concerned for the well-being of others is at the core of trust. This does not mean that the interests of others always come before the needs of the organization as a whole. However, it does require an understanding of the impact of one's actions on others and ways to balance the needs of individuals and the organization.

2.9   Defining organizational trust

The definition of *trust* varies from author to author. Some selected definitions of trust are highlighted in the following text. Zand (1997) defines trust as "a willingness to increase your vulnerability with another person, whose behavior you cannot control, in a situation in which your potential benefits are much less than your potential losses if the other person abuses your vulnerability." The vulnerability placed on an individual affirms that trust is a risk-based activity. Shaw (1997) defines trust as the belief that the individual being depended upon will meet expectations. Trust can be viewed as the willingness to rely or depend on some event, individual, group, or system. Trust requires a focus or an object of evaluation specific to the area of interest. According to Lane and Bachmann (1998), trust is a social phenomenon that makes accomplishing work within an organization easier and collaboration possible. Gilbert (1998) believes that organizational trust is the feeling of confidence in and support of an employer. Shockley et al. define organizational trust as "the organization's willingness, based upon its culture and communication behaviors in relationships and transactions, to be appropriately vulnerable based on the belief that another individual, group, or organization is *competent, open and honest, concerned, reliable,* and *identified* with common goals, norms, and values."

Many authors and theorists define trust based on elements or attributes that are needed to gain and sustain trust. These attributes or elements are oftentimes included in the definitions generated by these authors in defining trust. Trust has been defined frequently as ranging from having no specific attributes to having five attributes. Butler (1991) defined trust as having ten attributes that include availability, competence, consistency, discreteness, fairness, integrity/honesty, loyalty, openness, promise fulfillment, and receptivity. Trust is most frequently defined in terms of having at least one but not more than five elements or attributes as demonstrated in Table 2.3. The most frequently cited attributes of trust found in the literature are as follows: openness and honesty, competence, reliability, identification, and concern for employees.

2.10   Principal trust attributes

Collectively, authors and theorists believe that there are attributes of trust that are important in gaining and maintaining trust in organizations. The literature review highlighted as many as 12 trust attributes that are cited by authors and theorists as being important to building trust

Table 2.3 Trust Attributes Comparison

Author	Openness and honesty	Competence	Reliability	Concern	Identification	Availability	Consistency	Discreteness	Fairness	Loyalty	Promise fulfillment	Receptivity
Rotter, 1980	X											
Robbins, 1984	X											
Baiser, 1986		X		X								
Kouzes and Posner, 1987	X	X	X	X								
Mishra, 1992	X	X	X	X								
Fairholm, 1994	X	X	X									
Tway, 1994	X	X										
Mayer and Davis, 1995	X	X		X								
Kramer and Tyler, 1996	X											
Blomqvist, 1997		X	X									
Robbins, 1997	X	X	X									
Shaw, 1997				X								
Tozer, 1997	X											
Butler, 1991	X	X				X	X	X	X	X	X	
Tang and LiPing, 1998	X		X									
Reina and Reina, 1999	X	X										
Shockley-Zalabak et al., 1999	X	X	X	X	X							
Bennis, 2000	X	X	X	X						X	X	
Davis, Schoorman, Mayer, and Tan, 2000	X	X	X				X	X	X	X	X	X

(Continued)

Table 2.3 (Continued) Trust Attributes Comparison

Author	Openness and honesty	Competence	Reliability	Concern	Identification	Availability	Consistency	Discreteness	Fairness	Loyalty	Promise fulfillment	Receptivity
Tschannen-Moran, 2001	X	X	X	X								
Woolston, 2001	X	X			X							
Dotlich and Cairo, 2002	X	X										
Bracey, 2002	X		X	X								
Albrecht, 2002	X	X										

among people. These attributes include concern for employees, openness and honesty, availability, consistency, identification, reliability, discreteness, fairness, loyalty, promise fulfillment, receptivity, and competence. The five attributes featured in the study included openness and honesty, identification, competence, concern for employees, and reliability. These attributes are referred to as dimensions by Shockley et al. (1999), however, for the purpose of this study they are referred to as attributes based upon the varying ways trust has been defined by various authors. The five predominant attributes of trust identified through the literature search are necessary in order to build high-trust organizations.

The openness and honesty attributes are demonstrated by meeting commitments and promises. This attribute is the most frequently cited attribute by people when asked what attribute contributes most to organizational trust. The openness and honesty attribute involves the amount and accuracy of information that is shared and how sincerely and appropriately it is communicated. Trust makes performance easier, because it forms the basis for greater openness between individuals and groups. The attribute of openness and honesty accounts for more of the variance in believability than any other factor. If an individual is not viewed as being honest, his or her messages will not be accepted and will not be followed willingly.

The concern for employees attributes deals with the feeling of caring, empathy, tolerance, and safety demonstrated when people are vulnerable in business activities. According to Kouzes and Posner (1987), sensitivity to people's needs and interests is an important ingredient for building trust. It is important for managers to listen to what other people have to say and try to appreciate and understand their viewpoints. This act by management demonstrates respect for individuals and their ideas.

Reliability is concerned with whether an individual acts consistently and dependably. Simply put, can they be counted on to do what they say? One of the most important bases for an individual being perceived as reliable is predictability. Predictability refers to the degree of confidence that people have in their expectations with respect to another person's behavior or intentions.

The identification attribute refers to the extent to which groups or organizations hold common goals, norms, values, and beliefs associated with the organization's culture. Identification fosters commitment by shaping expectations about behaviors and intentions and leads to certain actions that will support the vision of an organization. Passion results from identification. Without identification there is no passion and very little, if any, commitment. A higher level of identification results in a higher level of commitment, loyalty, and performance.

The competence attribute deals with leadership competence. It does not specifically refer to the leaders' technical skills and abilities in the technical aspects of the business. Competence refers to qualities such as influence, ability, impact, expertness, knowledge, and the ability to do what is needed. Leaders are expected to display expertise in leadership skills that include the ability to challenge, inspire, enable, model, and encourage others to act in order to be viewed as being capable.

2.11   Summary of trust literature review

It is widely recognized that relationships are built on the premise of trust, and creating and maintaining an environment that fosters trust is not easily achieved. Successful development of products and processes requires a culture that is flexible, adaptive, utilizes participative decision making, empowerment, creativity, and teamwork. Organizational culture influences a firm's economic performance and plays an important role in shaping innovative processes. Cultures that facilitate innovation are credited with promoting teamwork, risk-taking, and creative actions needed in new technology development. These cultures are very important to the success of organizations. A culture characterized by trust provides the environment where positive interaction can occur and the work of the organization can be accomplished. Organizational trust establishes the framework needed for productivity. Trust creates an environment that encourages cooperation and permits employees to focus their attention on the tasks set before them. Organizational culture and trust are viewed as being of importance to a firm's success, because culture characterizes the environment in which work is performed and trust enhances the quality of decision making and the implementation of those decisions while stimulating productivity.

Result of empirical studies

Section II details the methods used to investigate the relationship between organizational culture and trust. The results of the two empirical studies are discussed in this section, including data analysis and reporting.

Section Two

Result of empirical studies

chapter three

Research statement and methodology

3.1 Introduction

This chapter describes the research question, the objective of the empirical study, and the hypotheses as they relate to studying the correlation between organic and mechanistic cultures and the level of trust found in organizations. Also discussed in this chapter is the methodology used to collect and analyze the data, survey selection methods, and the selection of the targeted population. The attributes of trust most frequently cited during the literature review are also investigated. These attributes include openness and honesty, concern for employees, competence, reliability, and identification.

3.2 Research question and objective

The review of the literature suggested that the answer to the following question could be a valuable enhancement to the body of knowledge for managers as they work to form successful high-performing organizations. The question is whether there a correlation between mechanistic and organic cultures and the level of trust found in technology-based organizations. The primary objective of this study was to test empirically the correlation between Burns and Stalker's mechanistic and organic cultures and trust in technology-based organizational environments. This research specifically investigated the correlation between employees' perception of trust and their perception of the culture within organizations, and in addition, tested the relationship between the attributes of trust identified through the literature review to determine their relationship with the organization's culture. The targeted population consisted primarily of technology-based organizations and their technical workers.

3.3 Hypotheses

This research explored six hypotheses:

1. H_{A0}: There is no correlation between culture (mechanistic and organic) and trust in technology-based organizations.
   H_{A1}: There is a correlation between culture (mechanistic and organic) and trust in technology-based organizations.

2. H_{B0}: There is no correlation between culture (mechanistic and organic) and the openness and honesty attribute of trust in technology-based organizations.

   H_{B1}: There is a correlation between culture (mechanistic and organic) and the openness and honesty attribute of trust in technology-based organizations.

3. H_{C0}: There is no correlation between culture (mechanistic and organic) and the competence attribute of trust in technology-based organizations.

   H_{C1}: There is a correlation between culture (mechanistic and organic) and the competence attribute of trust in technology-based organizations.

4. H_{D0}: There is no correlation between culture (mechanistic and organic) and the concern for the employee attribute of trust in technology-based organizations.

   H_{D1}: There is a correlation between culture (mechanistic and organic) and the concern for the employee attribute of trust in technology-based organizations.

5. H_{E0}: There is no correlation between culture (mechanistic and organic) and the identification attribute of trust in technology-based organizations.

   H_{E1}: There is a correlation between culture (mechanistic and organic) and the identification attribute of trust in technology-based organizations.

6. H_{F0}: There is no correlation between culture (mechanistic and organic) and the reliability attribute of trust in technology-based organizations.

   H_{F1}: There is a correlation between culture (mechanistic and organic) and the reliability attribute of trust in technology-based organizations.

3.4 Research significance and contribution to the body of knowledge

As an increasing number of technical workers move into management positions and as products and services increase in technology dependence, a clear understanding of the type of culture within organizations is important, inasmuch as culture affects trust, and trust affects performance. Having knowledge of the type of culture that produces high employee trust can be an important step in constructing the appropriate environment for building high-performing technology-based organizations, because trust is linked to a positive attitude among workers, higher levels of cooperation, and superior performance levels. Trust creates an environment that encourages cooperation and permits employees to focus their attention on the tasks set before them. The decision-making process can be affected by an individual's ability to trust, because trust enhances the quality of decision making and the implementation of those decisions, while stimulating productivity, because it gives people confidence that appropriate decisions are being made so that goals can be achieved.

This research is of value to managers as they prepare and shape high-tech organizations to compete successfully in a changing global environment. The findings can assist management in designing a culture that builds and maintains high trust levels thereby improving performance and the ability to compete successfully in an ever-changing global environment. This study is expected to provide an empirical and theoretical contribution to the literature attempting to provide a link between organizational culture and organizational trust. The most important contribution is that this study is the first to highlight empirically the role that mechanistic and organic cultures play in building and maintaining trust in technology-based organizations. One of the primary goals for organizational leaders should be to create a culture that will reinforce trust in order to position their organizations to compete successfully in their respective businesses.

3.5   Contribution to the disciplines of management and leadership

Cultures that foster trust are becoming more important to technology-based organizations as they continue to struggle to gain advantage in highly competitive markets. The culture of an organization can have a significant impact on the willingness to trust and is an important factor in determining the overall level of trust found in organizations. Because trust is linked to increased performance, it is important for managers to be aware of the type of culture that produces high trust. Managers should have a desire to create cultures that foster and reinforce trust because trust is an important catalyst for increased performance. Management should seek to encourage cultures that are compatible with trust so that performance will be enhanced. Thus, there is a need to understand the relationship between the culture of an organization and the level of trust present in that organization in order to optimize business outcomes.

The connection between culture and the five principal attributes of trust (openness and honesty, competence, concern for employees, reliability, and identification) can be utilized by management as tools to improve the level of trust in organizations. Investigating the relationship between organic and mechanistic cultures and the level of trust found in organizations is expected to lead to additional insights that can be used by managers to increase organizational performance. This is because it is recognized that trust influences employee performance and the sum of individual performances influences organizational performance as a whole.

3.6   Survey instrument selection

Survey instruments are commonly used to measure an individual's perception of various topics. Research using survey instruments is used to study populations by selecting and studying samples from the populations to discover the relative incidence, distribution, and interrelationships of sociological and physiological variables. Research using survey instruments has contributed greatly to the current knowledge base in the social sciences. Survey research has one unique advantage not shared by other social scientific methods. Namely, it is feasible, using relatively straightforward methods, to check the validity and reliability of survey data. Accordingly, survey instruments were used as a means to measure individual perceptions of organizational culture and trust during this study.

3.6.1   Measuring organizational culture

One important advantage of using survey instruments to study organizational culture is that the results obtained can serve as the basis for comparison and generalization of the data collected. The prevailing literature yielded a wide variety of definitions as well as methodologies used in studying culture. Culture has been extensively studied for many years, therefore there are many survey instruments available to measure it. These instruments differ in part due to variations in definitions and views of organizational culture.

In order to ensure selection of the appropriate survey instrument to measure culture, selection criteria were established. The criteria included selection of an instrument that (1) measured culture from organic to mechanistic, (2) was proven to be valid and reliable, (3) had a short completion time (<10 min), and (4) had a survey usage history. Two instruments were identified during the literature review that measure culture directly or indirectly on a scale from mechanistic to organic. These survey instruments included the Profile of Organizational Characteristics (POC) developed by Likert and the Organizational Culture Assessment (OCA) developed by Reigle. Both the POC and the OCA (2003) were determined to be valid based on previous studies. The POC survey instrument consists of 16 questions and the OCA survey instrument consists of 20 questions.

The POC survey instrument was developed to measure organizational characteristics that include leadership, motivation, communication, decisions, goals, and control mechanisms. The POC instrument was first validated in a 1964 General Motors study. The survey instrument was also shown to be valid and reliable by Reigle during validation of the OCA survey instrument. According to Taylor and Bowers (1972), the POC survey instrument has been extensively used to measure organizational variables. During recent years, others have used the POC survey instrument to measure culture on a scale from mechanistic to organic. The POC score for an organization is generated by determining the mean value for the completed surveys. The mean value is compared to the Likert System IV scale to determine where the organizational culture resides on the continuum. An organization that falls within the Likert System I structure can be viewed as possessing a mechanistic organizational culture, whereas an organization identified as having a Likert System IV management structure can be viewed as operating in an organic culture. Organizations falling in the Likert Systems II and III structures are viewed as having a culture in between mechanistic and organic on the culture continuum. A schematic representation of the Likert and Burns and Stalker systems and where each system resides on the cultural continuum is shown in Figure 3.1.

The OCA instrument was developed by Reigle as a means to measure culture on a scale from organic to mechanistic. The OCA was validated and found to be reliable using the data collected from high-technology organizations, but has not as yet been widely used. The instrument measures culture on a continuum scale from 1.0 to 8.0. The OCA score for an organization is generated by determining the mean values for each survey completed. Cultures displaying more organic characteristics will fall at the upper end and those displaying more mechanistic characteristics will fall at the lower end of the continuum. After tabulation of the scores, organizations receive an overall culture score that corresponds to a culture type ranging from Mechanistic (< = 4.74) to Organic (> 5.75).

The two instruments were evaluated based on the predetermined selection criteria. The result of the evaluation is shown in Table 3.1. The survey evaluation process consisted of survey reviews and interviews with participants who completed the culture surveys during the pilot study. These discussions provided validation that the questions posed by the survey instruments for the most part were easily understood and the surveys could be completed in a short period of time.

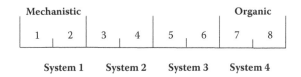

Figure 3.1 Likert and Burns and Stalker systems.

Table 3.1 Culture Survey Instruments Description

Instrument	Organic – mechanistic	Brief completion time	Easy to read and understand	No. survey questions	Usage history
POC (Likert, 1967)	Yes	Yes	Yes	16	Extensive
OCA (Reigle, 2003)	Yes	Yes	Yes	20	Limited

Upon completion of the screening process, both the POC and the OCA survey instruments met the criteria. However, it was noted that the OCA had not been extensively used to measure organizational culture. A limited pilot study was conducted using both the POC and the OCA survey instruments. Detailed results of the pilot study are discussed later in this chapter. The Cronbach Alpha reliability coefficient was calculated for both culture instrument datasets. Reliability demonstrates the extent to which a measurement can be repeated using the same measure of an attribute. The Cronbach Alphas for the POC and OCA data were 0.91 and 0.80, respectively. The POC survey instrument was selected to measure culture because it is widely recognized and well established and the instrument demonstrated a higher reliability coefficient during the pilot study for measuring culture.

3.6.2   *Measuring organizational trust*

Although the topic of trust has been widely researched during recent years, there have not been many instruments developed that specifically measure trust within an organization. In order to determine the appropriate survey instrument to measure organizational trust, the instrument selection process was based on four criteria. The criteria included: (1) an instrument that measures trust on a continuum scale, (2) ease of completion with limited completion time (<10 min.), (3) availability of reliability and validity data, and (4) the instrument's ability to measure multiple attributes of trust. The survey evaluation process consisted of survey reviews, interviews with participants who completed the culture surveys during the pilot study, and reviews of available reliability and validity data for each of the survey instruments evaluated. The discussions provided validation that the questions posed by the survey instruments for the most part were easily understood and the surveys could be completed in a short period of time. The instruments reviewed were evaluated based on the aforementioned criteria. These instruments were initially screened against the first three of the four criteria. This initial evaluation concluded that three of the instruments met the first three criteria,

Table 3.2 Trust Survey Instruments

Instrument	Brief completion time	Easy to read and understand	Valid and reliable
Conditions of Trust Inventory – CTI (Butler, 1991)	No	Yes	Yes
Intention to Trust Survey Instrument – ITSI (Clark and Payne, 1977)	Yes	Yes	Yes
Organizational Trust Index – OTI (Shockley et al., 1999)	Yes	Yes	Yes
Organizational Trust Questionnaire – OTQ (Chadwick, 2001)	No	Yes	Yes
Cornerstone Trust Survey – CTS (Birtel, Nellen, and Wilkes, 2004)	Yes	Yes	Yes
Interpersonal Trust Scale – ITS (Rotter, 1967)	No	No	Yes
Organizational Trust Scale – OTS (Garrity, 1991)	No	Yes	Yes
Management Behavior Climate Assessment – MBCA (Saskin and Levin, 2000)	No	Yes	Yes

as described in Table 3.2. These three instruments were the Cornerstone Trust Survey (CTS), the Intension to Trust Survey Instrument (ITSI), and the Organizational Trust Index (OTI).

The fourth screening criterion was used to evaluate the instruments' ability to measure the multiple attributes of trust identified through the literature review. The attributes of trust most often used in defining trust were selected for use in this research. These attributes are openness and honesty, competence, concern for employees, identification, and reliability. The ITSI did not measure any particular attribute of trust. The CTS survey measures three attributes of trust and the OTI measures five dimensions of trust. The attributes/dimensions of trust measured by each finalist is listed in Table 3.3.

The OTI survey instrument measures the dimensions/attributes of trust identified during the literature review. The OTI survey instrument was developed by Shockley et al. (1999) during a study funded by the International Association of Business Communicators (IABC) Research Foundation. The baselining process consisted of surveying approximately 4,000 employees encompassing 53 organizations across eight countries. The OTI was designed to measure organizational trust, while addressing five dimensions of trust on a continuous scale from 1 to 5. The survey shows trust results ranging from little trust (1) to very great trust (5).

Table 3.3 Survey Instrument Analysis

Instrument	Attributes/dimension of trust measured
ITSI	None
OTI	Openness and honesty
	Competence
	Concern for employees
	Identification
	Reliability
CTS	Competence
	Credibility (honesty)
	Care (concern)

The 29-question survey instrument was tested and validated for domestic and international use using LISREL 8 software to conduct confirmatory factor analysis. The result of the confirmatory factor analysis and structural equation modeling provided strong evidence that the OTI survey instrument measures five dimensions of trust that include concern for employees, openness and honesty, identification, competence, and reliability. The Cronbach Alpha reliability coefficient for the instrument is 0.95. The Alpha reliabilities for the five subscales, each of which measures one dimension of trust, ranged from 0.85 to 0.90. The OTI survey instrument was selected for use in this study.

3.7   Demographic questionnaire

A demographic questionnaire was developed to collect pertinent information needed to supplement the research initiatives. Each participant was asked to complete the demographic questionnaire. These demographic properties are expected to contribute to the study of employee trust within an organization by outlining whether the listed demographic characteristics have an impact on trust. The demographic questionnaire contained questions designed to answer the following:

- Organization name
- Work group name
- Individual job title
- Individual's level within the organization
- Organization type
- Organization size
- Individual gender
- Individual ethnicity
- Levels of management within the organization

- Age
- Time working for the organization
- Job tenure

3.8 Targeted population

The targeted sample population included knowledge- and technology-oriented workers who were employed by technology-based organizations, firms that compete in fields such as electronics, computers, data processing, information technology, chemicals, pharmaceuticals, communication, and instrumentation. These organizations typically employ a large number of technology-oriented workers. Technology-oriented workers are defined as workers who have advanced knowledge in the areas of science, engineering, and other technically related fields. Knowledge in these disciplines is generally acquired through formal education.

The survey instruments were administered to a sample of convenience instead of a random sample. The convenience sampling methodology has been extensively used and documented in various journals and dissertations. According to Jobber and Horgan (1988) the convenience sampling method is used extensively in research performed in the United States and Britain.

Technology-based organizations located in the southeastern United States were targeted for inclusion in the study due to accessibility. The firms that were targeted for participation in the study were located within a three-hour radius of each other for researcher travel feasibility and to allow economic data collection.

3.9 Data collection process

The data collection process proceeded using the culture and trust survey instruments and the demographic questionnaire. The data-gathering process for each organization consisted of meetings and presentations to management or management teams. Survey instruments were delivered to management for distribution. Upon completion, the participants placed the completed surveys in the envelopes provided and sealed them. The sealed envelopes containing the completed surveys were collected from management in most cases. In one instance, the surveys were delivered and returned through the postal mailing system due to the travel time to and from the participating company.

chapter four

Pilot study

4.1 Conducting the pilot study

A pilot study was performed prior to embarking upon the full-scale study to assist in determining the viability of the proposed research methodology. The data were collected using the POC survey instruments to measure culture, the OTI survey instrument to measure trust, and the demographic questionnaire. The sample consisted of 51 technical professionals across three (3) organizations. The three organizations that participated in the study are referred to as Organizations A, B, and C. The collected data were analyzed using Pearson correlation analysis to determine the relationship between culture and trust. The mean culture and trust results were also calculated and reviewed.

4.1.1 Organization A

Organization A operates in a research environment providing engineering services to a variety of projects serving both government as well as the private business sector. This organization is recognized as playing an important role in our nation's security. The sample from this organization did not contain any minority representation. The demographic data for the organization are shown in Table 4.1.

4.1.2 Organization B

Organization B is a government organization that provides support to various government and public initiatives across the country, as needed. The organization is governed by extensive use of procedures and has processes in place to ensure that the chain of command is followed. The demographic data for Organization B are listed in Table 4.2.

4.1.3 Organization C

Organization C provides a variety of consultation services to various projects and organizations. Some of the consultation support includes setting policies and developing procedures along with providing various laboratory analytical procedures. The demographic data for Organization C are listed in Table 4.3.

Table 4.1 Organization A Sample Demographics

Demographics	% Population
Females	37
Males	63
African Americans	0
Caucasians	100
Management	10.5
Nonmanagement	89.5

Table 4.2 Organization B Sample Demographics

Demographics	% Population
Females	22.2
Males	72.2
African Americans	72.2
Caucasians	27.8
Management	15.8
Nonmanagement	84.2

Table 4.3 Organization C Sample Demographics

Demographics	% Population
Females	42.1
Males	57.9
African Americans	26.3
Caucasians	73.7
Management	94.7
Nonmanagement	5.3

4.2   Pilot results

A total of 51 data points was collected and used in the pilot study. Normality analysis was performed for both the organizational trust and organizational culture data. The normality analysis (Figures 4.1 and 4.2) demonstrated that the data collected were representative of an approximate normal population.

A correlation analysis was performed using the Pearson correlation method to determine the correlation between trust and culture for the entire dataset. The correlation coefficient for the organizational trust and culture data was 0.55 with a *p*-value of <0.001. It is recognized that correlation coefficients between 0.30 and 0.70 are significant in determining

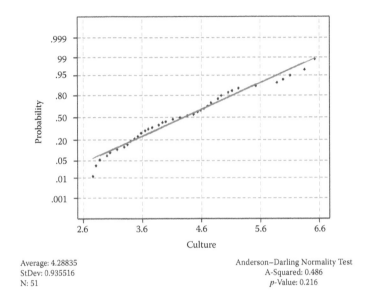

Figure 4.1 Normality plot for culture data.

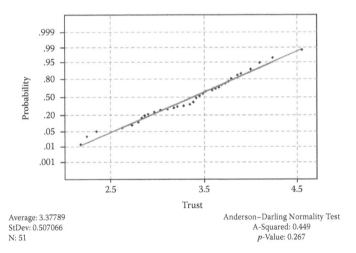

Figure 4.2 Normality plot for trust data.

the relationship between two datasets. The correlation coefficient for the limited pilot study demonstrates that a correlation between organizational culture and trust exists.

The mean for the organizational trust survey and the culture survey were 3.4 and 4.2, respectively. The mean trust scores for the organizations ranged from 3.2 to 3.5 and the mean culture scores ranged from 3.9 to 5.0. Organization A is noted as having both the lowest culture and trust means of 3.9 and 3.2, respectively. Organization C was recorded as having the highest culture and trust means of 5.0 and 3.5, respectively. The culture and trust means for each of the organizations further demonstrates that there is a relationship between organizational culture and organizational trust. The culture and trust means for each of the organizations are summarized in Table 4.4 and shown in Figure 4.3.

The bar chart in Figure 4.3 provides a pictorial view of the data. The chart shows that as the culture scores increase the trust scores also increase and as the culture scores decrease the trust scores decrease.

Table 4.4 Culture and Trust Means—Pilot

Organization	Trust	Culture
A	3.2	3.9
B	3.3	4.5
C	3.5	5.0

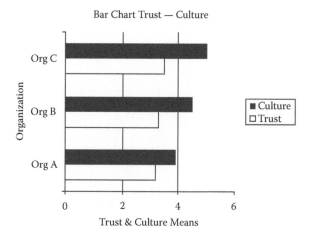

Figure 4.3 Bar chart for culture and trust pilot data.

4.3   Summary of pilot study

The pilot study results support the premise that the culture of an organization can be important in terms of building and maintaining trust. Specifically, the results show that there is a positive correlation between organizational culture and organizational trust. The demographic data show a difference in trust levels for males and females. There was also a difference noted in trust levels for African Americans and Caucasians. Because the sample size was small, additional data collection was needed to further validate the theory and findings.

4.4   Lesson learned

The most important lesson learned as a result of the pilot study involved confidentiality of information. Survey participants were concerned with the possibility of being connected to their completed surveys in some way. Therefore, some potential participants were reluctant to complete the surveys. Some employees refused participation, fearing that the information they were asked to provide might be traceable to them. As a result of the confidentiality concerns encountered during the pilot, modifications to the survey distribution and collection process were made. During the research, surveys were distributed in envelopes with instructions for the survey participants to seal them after completion. In addition, a statement assuring survey participants that confidentiality of information would be maintained was added to each envelope.

4.3 Summary of pilot study

4.4 Lesson format

chapter five

Full study

5.1   Introduction

This chapter documents the data analysis process and the results of the full-scale study to determine if a correlation exists between mechanistic and organic cultures and the level of trust found in technology-based organizations. The data collection process extended over a one-year period and is consistent with the process used during the pilot study, taking into consideration the lessons learned from that study. A total of 608 data points were collected and used in the study. Each participating organization having a representative sample was included in a separate organizational analysis that is discussed later in the chapter. Data were collected using the culture and trust survey instruments tested during the pilot study, the Likert Profile of Organizational Characteristics (POC) and the Organizational Trust Index (OTI). A demographic section was also used to collect pertinent demographic information needed to further characterize and validate the data.

5.2   Description of data analysis methods

The data analysis process consisted of basic statistical methods. These methods were used to evaluate the applicable hypotheses. The statistical methods included the Anderson Darling Normality Test, the Wilcoxon Sign Test, sample means comparison, and correlation analysis, probability analysis, Tukey's Range Test, and survey reliability analysis. These methods were used to compare and determine the relationship between variables, to describe the central tendency for the dataset, express the frequency of occurrence of an event, determine repeatability of measurement, and correlation of the data.

5.3   Data characteristics

The population gender data consisted predominately of male respondents (65.1%) and the predominant average age group of the sample population was between 46 and 55 (31.3%). Most of the sample consisted of government contract workers (47%) versus government (26.4%) and private industry (26.7%). Government contract workers are defined as workers who are employed by an organization that is under contract with the government to work on projects administered by the government. A detailed analysis of pertinent demographic data is discussed later in this section. A summary of the data collected can be found in Tables 5.1 to 5.6.

Table 5.1 Ethnic Background

Ethnicity	Number	% Population
African American	146	24.0
Caucasian	422	69.3
Hispanic	9	1.5
American Indian	4	0.7
Asian	4	0.7
Other	5	0.8
No response	18	3.0

Table 5.2 Organization Tenure (Years)

Tenure	Number	% Population
<1 year	32	5.3
1 to <5	108	17.8
5 to <10	80	13.2
10 to <15	61	10.0
15 to <20	143	23.5
20 to <30	120	19.7
30 or more	14	2.3
No response	50	8.2

Table 5.3 Age

Age	Number	% Population
<25	45	7.4
25 to 35	98	16.1
36 to 45	151	24.8
46 to 55	190	31.3
55+	65	10.7
No response	59	9.7

Table 5.4 Organization Type

Org. Type	Number	% Population
Government	160	26.4
Government Contractor	286	47.0
Private Industry	162	26.6

Table 5.5 Position in Organization

Position	Number	% Population
Management	110	18.1
Nonmanagement	465	76.5
No response	33	5.4

Table 5.6 Gender

Gender	Number	% Population
Male	396	65.1
Female	206	33.9
No response	6	1.0

5.4 *Participating organizations*

The organizations that participated in the study depend greatly upon their knowledge of and advances in technology to compete and perform work in their respective markets, and to serve their customers' needs and expectations successfully. Data were collected from a total of 17 organizations. However, only 10 of the organizations represented by the study are included in the organizational analysis due to small sample sizes for seven of the organizations. Three of the organizations that were included in the organizational analysis were from the private sector, four of the organizations were governmental agencies, and three were government contracting organizations.

A total of 608 data points was collected. Forty-seven data points were not included in the organizational analysis because the sample sizes from the organizations represented were small (less than 10). Only those organizations with enough data points presumed to be representative of the organization based upon random distribution of surveys were included in the organizational analysis. The 47 data points not included in the organizational analyses were also collected from government, government contract, or private industries. These data points were included in the aggregate data analysis. The organizations participating in the study are referred to as organizations D to M. Each organization is discussed briefly in Table 5.7. The complete demographic data for each

Table 5.7 Sample Collected by Organization Data

Organization	No. samples collected	Organization type	Size of organization
D	70	Government	>1,000
E	15	Government contractor	<1,000
F	257	Government contractor	>1,000
G	35	Government	>1,000
H	23	Private	>1,000
I	15	Government	>1,000
J	85	Private	100–250
K	15	Private	>1,000
L	11	Government contractor	>1,000
M	35	Government	<100
Misc.	47	Government, government contractor, private	NA

of the organizations that participated in the full study are shown in the appendix.

5.4.1 Organization D

Organization D provides support to a diverse segment of customers that include government as well as nongovernment clients in the area of communication and intelligence technologies. Survey instruments were completed primarily by employees from four groups resulting in a total of 70 completed surveys. The size of each group varied due to the nature of the work performed. Employees working in this organization are accustomed to complying with a variety of rules and procedures.

5.4.2 Organization E

Organization E plays an important role in the research community and in providing technological advancements for a diverse customer base. This organization employs a large number of engineers and scientists who engage in a large variety of projects using cutting-edge technologies. Data were collected primarily from one group resulting in 15 completed surveys. The sample collected from this organization is considered to be relatively small when compared to the actual size of the organization.

5.4.3 Organization F

Organization F provides a variety of support and technological services in various technology development and research initiatives. In many cases, the service provided involves the use and development of cutting-edge technologies and processes. Data were collected primarily from employees of five groups resulting in a total of 257 completed surveys.

5.4.4   Organization G

Organization G provides services to a wide range of clients. This organization depends greatly on advances in science and technology in order to provide quality services to a diverse client base. Survey instruments were completed primarily by employees from three groups resulting in a total of 35 completed surveys. The groups surveyed consisted primarily of engineers and technicians. The average size of each group in the organization varied depending on the group's specific mission or function.

5.4.5   Organization H

Organization H is a large privately owned organization with employees located in multiple locations throughout the southern United States. This organization provides services to a large diverse customer base. Some of the employees in this organization at times do not feel connected to the organization and top management due to the distance between the group and the corporate office. The average size of the work groups within Organization H was 20 to 45 employees. Data were collected primarily from employees from one work group resulting in a total of 23 completed surveys.

5.4.6   Organization I

Organization I is a government owned and operated entity that utilizes a highly skilled technical workforce to ensure the goals of the organization are successfully achieved. The workforce consists primarily of engineers and scientists who are specialists in their respective fields. Data were collected from employees from one work group resulting in a total of 15 completed surveys representing approximately 50% of the workgroup. The sample collected from this organization is considered to be relatively small when compared to the actual size of the organization.

5.4.7   Organization J

Organization J is a privately owned company located in the southern area of the United States. This company is dedicated to producing and providing specialized products to a large customer base. Most of the management team members are technically skilled and in some cases had little or no management or soft skills training. Data were collected from each work group resulting in a total of 85 completed surveys.

5.4.8   Organization K

Organization K competes in a market that is beginning to expand. Only one group within Organization K was sampled resulting in 15 completed surveys. The sample collected from this organization is considered to be relatively small when compared to the actual size of the organization. This group consisted primarily of engineers and technicians.

5.4.9   Organization L

Organization L provides a variety of technical support to a large customer base. Support includes engineering, computer systems design, tactical communication, program management, and system modeling and simulation. This organization also participates in a variety of tactical systems development using a variety of technologies. A total of 11 surveys was completed by one group. The sample collected from this organization is considered to be relatively small when compared to the actual size of the organization.

5.4.10   Organization M

Organization M is an organization that engages in a variety of research activities. The researchers are organized into four groups. Data were collected from each of the four groups resulting in a total of 35 completed surveys. The workforce is highly educated and skilled in various scientific and technical disciplines.

5.5   Data and analysis of results

The data collected during the study were analyzed with the assistance of two statistical software packages. The Minitab and the SPSS statistical software programs designed for statistical data analysis were used. Actual data analysis consisted primarily of calculating the mean for each dataset, normality, Wilcoxon Sign Test, Tukey's Test, reliability analysis, and correlation analysis. The data were analyzed to test the following hypotheses.

H_{A0}: There is no correlation between culture (mechanistic and organic) and trust in technology-based organizations.

H_{A1}: There is a correlation between culture (mechanistic and organic) and trust in technology-based organizations.

5.6 Normality and survey reliability analysis

A normality analysis was performed for both the organizational culture and trust data using the Anderson Darling Normality Test. The analysis yielded a *p*-value of <0.001 for the trust data and a *p*-value of 0.021 for the culture data. The normality plots for the data are shown in Figures 5.1 and 5.2.

A reliability analysis was performed for both the culture and trust data collected. The alpha reliabilities for the data collected using the culture and trust instruments were 0.95 and 0.97, respectively. The results indicated that the data collected using the two survey instruments can be repeated when surveys are completed by respondents of the same organizations. The alpha reliability was also calculated for each of the trust attributes. The reliability coefficients ranged from 0.85 to 0.93. The reliability result for each attribute of trust is listed in Table 5.8.

5.7 Mean analysis

The mean for the culture data was 4.72 and the trust data was 3.49. An OTI score of 3.49 indicates that the respondents have some trust in their respective organizations. A culture result of 4.72 demonstrates that the culture overall is not entirely organic or mechanistic. The culture score for the sample indicates that the respondents perceived that their

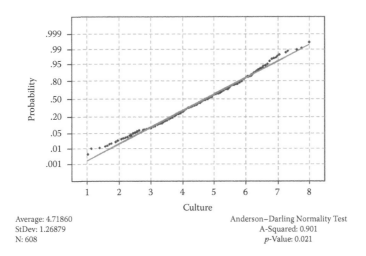

Average: 4.71860
StDev: 1.26879
N: 608

Anderson–Darling Normality Test
A-Squared: 0.901
p-Value: 0.021

Figure 5.1 Normality plot—culture data.

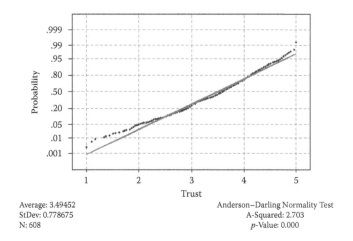

Average: 3.49452
StDev: 0.778675
N: 608

Anderson–Darling Normality Test
A-Squared: 2.703
p-Value: 0.000

Figure 5.2 Normality plot—trust data.

Table 5.8 Reliability Coefficient—Trust Attributes

Attribute	Reliability coefficient
Competence	0.88
Openness and Honesty	0.93
Concern for Employees	0.89
Reliability	0.85
Identification	0.85

work culture is between organic and mechanistic on the culture continuum. The trust attribute means resulting from this study are shown in Table 5.9.

A review of the attribute means for the entire dataset show the highest mean for the reliability attribute followed by the identification and concern for employee attribute. The lowest mean was calculated for the competence attribute as shown in Figure 5.3.

The OTI scores for the organizations reported for each attribute of trust show that the attribute of reliability was the highest for five of the ten organizations that participated in the full study. The identification and competence attribute received the highest mean for two of the organizations surveyed. This result may indicate that for the organizations participating in the study, having a reliable leadership team is of most importance to the workers. The trust attribute means for each organization are listed in Table 5.10 and Figures 5.3 to 5.13.

Table 5.9 Trust Attributes

Trust attributes	Trust means
Competence	3.43
Openness and Honesty	3.45
Concern for Employees	3.50
Reliability	3.63
Identification	3.50

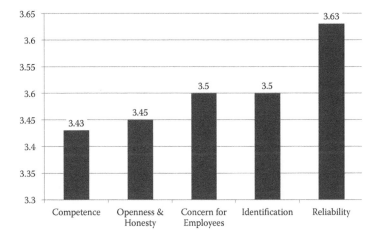

Figure 5.3 Trust attribute means.

Table 5.10 Trust Attributes Means

Organization	Openness and honesty	Concern for employees	Identification	Competence	Reliability
D	3.18	3.31	3.20	2.96	**3.42**
E	3.64	3.6	**3.77**	3.57	3.72
F	3.57	3.62	3.66	3.54	**3.75**
G	2.77	2.60	**2.78**	2.66	2.66
H	3.79	3.65	**3.97**	3.66	3.95
I	3.78	3.95	3.64	3.6	**4.0**
J	3.20	3.19	3.22	**3.47**	**3.47**
K	3.74	3.81	3.75	**3.87**	3.82
L	3.65	3.56	3.53	3.43	**3.68**
M	3.73	**3.77**	3.55	3.72	3.7

Figure 5.4 Organization D trust attribute means.

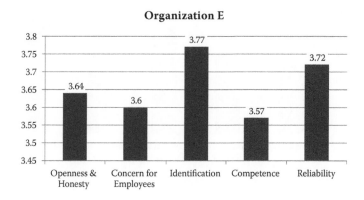

Figure 5.5 Organization E trust attribute means.

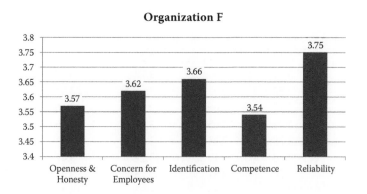

Figure 5.6 Organization F trust attribute means.

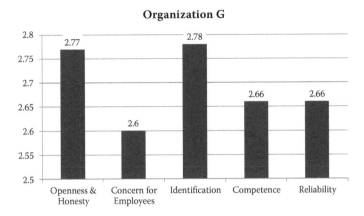

Figure 5.7 Organization G trust attribute means.

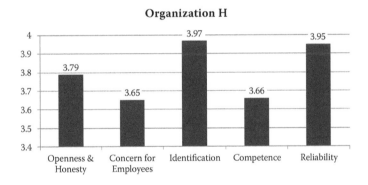

Figure 5.8 Organization H trust attribute means.

Figure 5.9 Organization I trust attribute means.

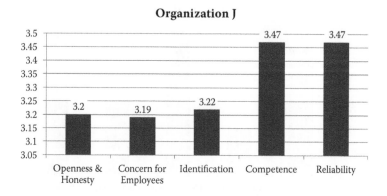

Figure 5.10 Organization J trust attribute means.

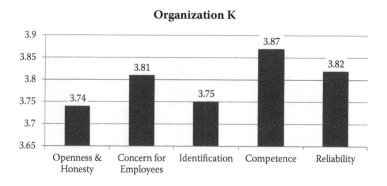

Figure 5.11 Organization K trust attribute means.

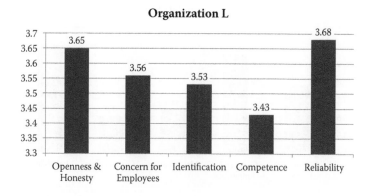

Figure 5.12 Organization L trust attribute means.

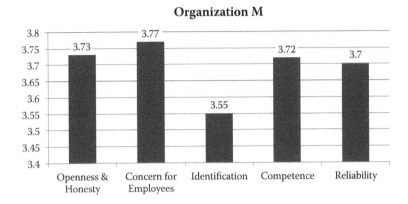

Figure 5.13 Organization M trust attribute means.

5.8 Correlation analysis

The scatterplot of the data collected for both culture and trust was evaluated and a correlation analysis was performed to determine the extent of correlation between culture and trust for the data collected. The Spearman Rho (SR) correlation coefficient was found to be 0.79 at a significance of 5%. The correlation result indicated a strong positive correlation between culture and trust, demonstrating that culture and trust are strongly linked and changes in culture can affect trust among people and in organizations.

5.9 Hypothesis testing—trust attributes

An analysis was performed to determine if there is a correlation between the trust attributes and culture. The hypotheses for the attributes of trust that were evaluated are listed below:

H_{B0}: There is no correlation between culture (mechanistic and organic) and the individual trust attributes in technology-based organizations.
H_{B1}: There is a correlation between culture (mechanistic and organic) and the individual attributes in technology-based organizations.

A correlation analysis was performed to determine if a positive correlation between culture and the attributes of trust existed for the data collected. The Spearman Rho correlation analysis demonstrated that there was a significant correlation between the trust attributes and culture. Correlation results and *p*-values are listed in Table 5.11.

Table 5.11 Trust Attributes Correlation with Culture and Significance Level

Trust dimension	Correlation coefficient	P-value
Openness and honesty	0.76	<0.001
Competence	0.70	<0.001
Concern for employees	0.73	<0.001
Identification	0.68	<0.001
Reliability	0.68	<0.001

The openness and honesty attribute correlated highest with a correlation coefficient of 0.76. Correlations of 0.70 or above are generally considered highly correlated. This attribute is believed to be highly important in the trust-building process. The resultant p-values of <0.001 were compared to a significance level of 0.05 and indicated that the null hypotheses can be rejected while accepting the alternative hypotheses demonstrating that there is a correlation between the trust attributes and culture.

5.10   Analysis by organization

An organizational analysis was conducted for organizations participating in the study that had adequate survey representation. As a result, 10 organizations were included in the organizational analysis. The sample size for the organizations ranged from 11 to 257 survey participants. The results of the organization analysis show that there is a positive correlation between culture and trust for each organization. The results also show that the more organic a culture was perceived, the higher the level of trust that was present, whereas, the more mechanistic a culture was perceived, the lower the level of trust (Figure 5.14).

The trust mean for the organizations ranged from 2.7 to 3.8 indicating trust levels from some to great trust. The culture means for the organizations ranged from 3.52 to 5.5. The organization with the smallest OTI (2.70) and culture (3.52) scores is Organization G. A lower trust result was noted among the organizations when the culture was perceived to be more mechanistic. Organization E trust and culture scores were among the highest at 3.8 and 5.7, respectively. A higher trust result was noted among the organizations that participated in the study when the culture was perceived to be more organic. The mean culture and trust results for each organization along with the correlation coefficients are shown in Table 5.12. The organizations are listed in order of trust results.

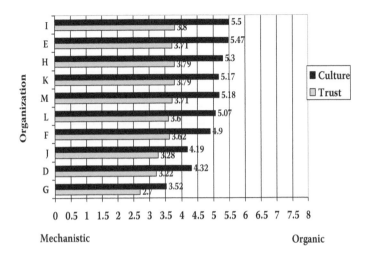

Figure 5.14 Bar graph for culture and trust—organizational means.

Table 5.12 Means by Organization Data

Organization	Trust mean	Culture mean	Correlation coefficient
G	2.70	3.52	0.85
D	3.22	4.32	0.70
J	3.28	4.19	0.74
F	3.62	4.90	0.79
L	3.60	5.07	0.53
M	3.71	5.18	0.79
E	3.71	5.47	0.49
K	3.79	5.17	0.88
H	3.79	5.30	0.57
I	3.80	5.50	0.71

5.11 Summary of organizational analyses

The culture means for each organization fell on the culture continuum between the mechanistic and organic regions ranging from 1.0 (mechanistic) to 8.0 (organic). In those organizations included in the organizational analysis, culture results ranged from 3.52 to 5.50. The organization with the lowest culture and trust means (organization G) displayed predominately mechanistic characteristics such as close adherence to the chain of command, reliance on standards and policies, and specialized job tasks. The organization with the highest culture and trust means (organization E) displayed

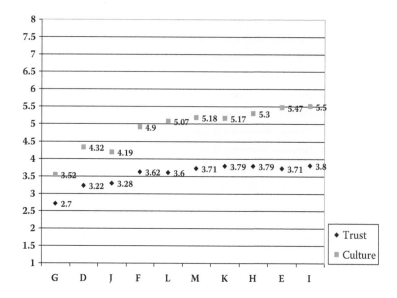

Figure 5.15 Plot for culture and trust—organizational means.

several organic characteristics such as team decisions made predominately by the individuals with the most task knowledge; communication was plentiful and flowed in all directions across the organization. The characteristics of these organizations were consistent with the findings of this study demonstrating that there is a relationship between organizational culture and the level of trust in organizations. The result seems to further suggest that the more organic a culture is perceived, the higher the levels of trust that are expected to be present. This relationship is shown in Figure 5.15. The correlation between the trust means and the culture means for each organization was 0.94 with a *p*-value of <0.001. The correlation results further demonstrate that there is a relationship between the culture of an organization and the level of trust found in that organization.

5.12   Trust in technology-driven organizations

Trust results for the organizations ranged from some to great trust. The trust scores were rounded to whole numbers prior to comparison with the trust index as the index reports trust levels in whole numbers. These organizations are competing successful in their various markets with trust levels ranging from some to great. However, productivity levels for the organizations are not known as this variable was not included in the study. Although not addressed by this study, one would conclude that the organizations exhibiting a great trust level would be expected to have a greater productivity level than the organizations with some trust level.

Table 5.13 Trust Levels by Organization

Organization	Trust level
G	Some
D	Some
J	Some
F	Great
L	Great
M	Great
K	Great
H	Great
E	Great
I	Great

Further exploration of this relationship has the potential to lead to further insight into the importance of trusting cultures. Trust levels are listed in Table 5.13 for the participating organizations.

5.13   Gender analysis

Studies on the communication patterns of men and women indicate that there is a difference in the way men and women communicate. The literature consistently reports differences between the work attitudes of men and women. For example, women are said to be more committed, more empathetic, and display less cynicism than men. Men are viewed as dominant, aggressive, competent, and persuasive in communication; conversely, women are viewed as passive and cooperative in communication. It is believed that women generally choose language that is considered nurturing, expressive, and supportive, whereas men generally use language that is considered opinionated and reactive.

The gender analysis demonstrated that overall the women and men participating in the study did not respond the same to the survey questions. Higher trust levels were noted most among the male respondents (Table 5.14 and Figure 5.16) for the organizational analysis. The trust mean for the overall dataset (including the data not included in the organizational analysis) showed the trust means for females at 3.47 and 3.51 for males. The Wilcoxon Sign Rank Test was used to determine if a difference existed between trust levels for males and females. A resultant p-value of 0.006 was compared to a significance level of 0.05 indicating that overall, there is a difference in trust levels for the males and females who participated in the study. A review of the gender analysis comparison for each trust attribute for the entire dataset show (Figure 5.17) that the reliability attribute is highest for the attributes for both males and females.

Table 5.14 Gender Analysis

Organization	Trust females	Trust males
D	3.00	3.27
E	3.68	3.73
F	3.68	3.61
G	2.51	2.93
H	3.70	4.76
I	3.76	3.96
J	2.96	3.33
K	3.05	3.90
L	3.39	3.77
M	3.72	3.70

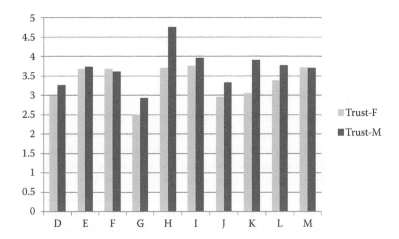

Figure 5.16 Gender trust by organization.

5.14   Ethnic group analysis

The trust scores were calculated for each ethnic group for each of the organizations. The results demonstrated that differences in ethnicity showed some differences between the trust results found in organizations. It must be noted that the sample sizes for Hispanics, American Indians, and Asians were small or nonexistent for most of the organizations that participated in the study. The population consisted of 69.4% Caucasians, 24% African Americans, 1.5% Hispanics, and 0.7% Asians and American Indians. Trust means were calculated for the ethnic groups represented by the study. Highest trust levels were found to be among American Indians (3.74) and Asians (3.8) followed by African Americans (3.65)

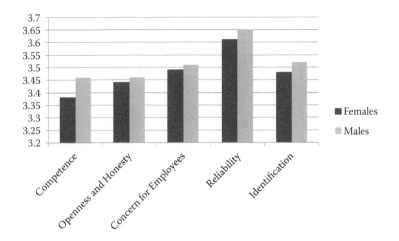

Figure 5.17 Trust attributes by gender.

Table 5.15 Ethnical Group Analysis

Ethnical group	African American	Caucasian	Hispanic	Asian	American Indian
Number	146	422	9	4	4
Mean	3.65	3.53	3.42	3.80	3.74
SD	0.877	0.732	0.660	0.572	0.411

workers. Inspection of the data for ethnicity demonstrated incremental differences in level of trust among races for the entire dataset. The results for the survey participants are included in Table 5.15 and Figure 5.18. See Figure 5.19. Additional review of the data for Caucasian and African Americans was conducted inasmuch as most of the sample came from the two groups. The Wilcoxon Sign Rank Test was performed to determine if a difference existed between trust levels for Caucasian and African Americans. The resultant p-value of 0.009 demonstrates that there is a difference between trust levels for the two groups.

Although most authors believe that the openness and honesty attribute of trust is most important in building trust, in observing the trust mean results for the trust attributes for each ethnic group completing the survey, it shows that reliability appears to be the most important attribute among the groups. Intuitively one may agree that honesty is of greater importance than any other attribute based on traditional belief. However, the result for this study highlighted a different attribute. The results may indicate that technical workers value working in an environment where the management team and the organization are reliable in business dealings and actions.

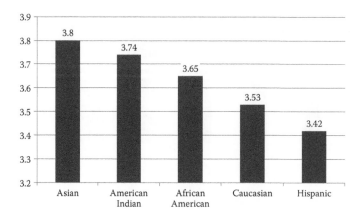

Figure 5.18 Ethnicity trust means.

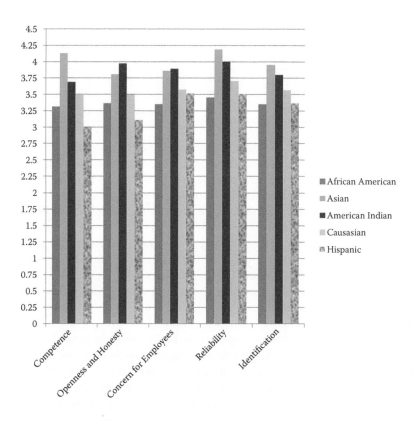

Figure 5.19 Ethnicity.

5.15   Age analysis

It is widely recognized that many workers are retiring at a later age than in previous years. Most forward-thinking companies have come to the realization that the future success of the organization can depend on the important role that mid-career and older workers play in the organization. The age data categories are shown in Table 5.16. Tukey's Test was performed to determine trust significance between the age group categories. The results demonstrated that trust levels differed for employees who were less than 25 years of age, between the ages of 46 and 55, and employees who were above 55 years of age. Trust levels were the highest for workers above the age of 55 and lowest for workers below the age of 25.

Upon inspection of the means for each attribute, it was noted that the reliability attribute mean result was slightly higher for each group (Figure 5.20).

Table 5.16 Age Analysis

Age group	<25	25–35	36–45	46–55	55+
Number	45	100	151	189	65
Mean	2.90	3.55	3.52	3.62	3.68
SD	0.801	0.792	0.777	0.757	0.750

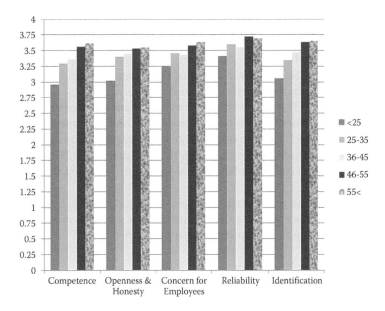

Figure 5.20 Age—trust attributes.

Table 5.17 Job Tenure Analysis

Tenure	<1	1<5	5<10	10<15	15<20	20<30	30+
Number	33	107	81	61	142	120	14
Mean	3.41	3.38	3.57	3.48	3.69	3.60	3.69
SD	0.889	0.754	0.929	0.722	0.752	0.663	0.865

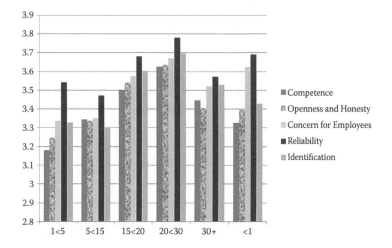

Figure 5.21 Job tenure—trust attributes.

5.16   Job tenure analysis

Some theorists believe that job tenure is related to employee trust in management and report lower levels of trust among employees as organizational tenure increases. Job tenure data characteristics are shown in Table 5.17. Tukey's Test was performed to determine the trust level significance between the tenure group categories. The results demonstrated that trust levels differed for employees who were employed by an organization for 1 to less than 5 years, 5 to less than 10 years, and 20 to less than 30 years. The result overall indicated that job tenure is related to trust levels in organizations.

A review of the trust attribute analysis for job tenure shows the reliability attribute once again showing the highest mean level among all of the attributes for each job tenure category (Figure 5.21).

5.17   Industry analysis

The data were analyzed based on the type of industry for which the survey respondents performed work. The Wilcoxon Sign Rank Test was used to determine if a difference exists between trust levels for the industries

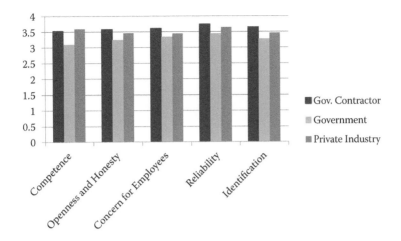

Figure 5.22 Industry type.

participating in the study. The results overall (p-value <0.001) demonstrate that there is a difference between the trust levels for employees working in the industries evaluated. The OTI score for government workers was 3.28 and the OTI score for private industry workers and government contract workers was 3.62. The result for the industry analysis is in line with what one would expect because it is generally recognized that government workers are oftentimes working in an environment where more mechanistic attributes are an integral part of the culture. An analysis of the trust development for each industry type is shown in Figure 5.22. The results show that reliability attribute trust means were highest for all industry types.

5.18   Job status/position analysis

Many researchers are of the belief that employees who are among the management ranks are more likely to have higher trust in their organizations as opposed to nonmanagement employees. People with management or supervisory responsibilities are more likely to align themselves closely with the values of the organization. The sharing of organizational values can have a positive impact on trust in organizations. An evaluation of trust levels for management and nonmanagement workers was performed using the Wilcoxon Sign Rank Test to determine if there was a difference between the trust levels for the two groups. The results (p-value 0.014) showed that overall there is a difference in trust levels among the two groups. The trust mean result for the entire dataset (including the data not included in the organizational analysis) showed mean results of 3.7 for management and 3.44 for nonmanagement workers. The means

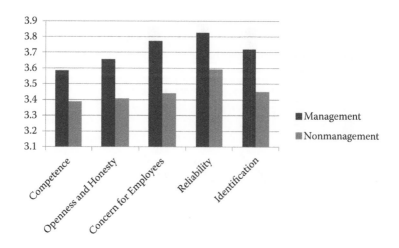

Figure 5.23 Trust attribute—job status.

for each attribute of trust are shown in Figure 5.23. The reliability attribute for management and nonmanagement showed the highest means.

5.19   Conclusion

The result of the analysis demonstrated that the culture of an organization does have an impact on the level of trust found in organizations. A review of the correlation result yielded a correlation coefficient of 0.79 indicating a strong correlation between culture and trust. The normality analysis conducted using the Anderson Darling Normality Test showed that the data were collected from an approximately normal population. Reliability analysis for the data collected using the trust and culture survey instrument were 0.97 and 0.95, respectively.

The results of the trust attribute analysis showed a strong correlation between culture and the attributes of trust. Correlation results ranged from 0.68 for the reliability attribute to 0.76 for the openness and honesty attribute. It was consistently noted during the literature review that the attribute of openness and honesty was viewed theoretically as the most important attribute of trust. Although the study confirmed the importance of being open and honest in communication if trust is to be gained and maintained, the study also showed that reliability is just as or, in most cases, more important. The result may suggest that for technology or knowledge workers, a reliable management team and organization are among the important characteristics of a desired work environment.

The cultures exhibited by the participating organizations were between mechanistic and organic on the culture continuum. The culture result obtained from the study was consistent with the culture results obtained by Reigle during her research while examining the cultures of 16 organizations. The culture means for each of the 16 organizations that participated in the Reigle study ranged from 3.7 to 5.9. The culture results for the organizations participating in this study ranged from 3.5 to 5.5. This consistency in results for the two studies may support a premise that the culture required for many organizations to compete successfully in a global economy consists of a combination culture containing the appropriate balance of both mechanistic and organic characteristics.

Analysis of specific demographic characteristics such as ethnicity, gender, position status, age, and job tenure showed that demographics did play a role in the level of trust within the organizations. Overall, the result of the study showed that there is a strong correlation between organizational culture and organizational trust in technology-based organizations. The result of the study also showed that the more organic the culture of an organization is perceived, the higher the level of trust present.

chapter six

Technology-driven organization cultures

6.1 Characteristics of technology-driven organization cultures

It is clear from the study that technology-based organization cultures are not completely organic or mechanistic in nature. Therefore, any attempt to define these cultures in past traditional terms is misleading to say the least. In order to compete successfully, technology-driven companies must pay attention to the prevailing culture. When culture is neglected or ignored, the company is sure to experience some type of negative impact such as reduced productivity. A quick summary of the literature review and the survey results indicate the following:

- Trusting culture is important to organization growth.
- The culture of an organization directly affects the level of trust found in the organization.
- Productivity is affected by trust.
- Technology-driven organization cultures are not purely organic or mechanistic.

Based on the study and the result of the literature, in purely mechanistic cultures one would expect that an organization would exhibit low trust; conversely in purely organic cultures the level of trust would be expected to be considerably higher. Considering the result of the study, it seems appropriate to ascertain that technology-driven organizational cultures are not completely organic or mechanistic. Cultures that permit flexibility, creativity, critical thinking, and the appropriate level of risk taking are viewed as optimal for technology-driven organizations. Also, noted among the organizations that participated in the study, each organization exhibited various characteristics or elements of organic and mechanistic culture. This information suggests that technical organizations function primarily in cultures that contain the right balance of attributes permitting critical thinking and flexibility and the right balance of elements supporting the appropriate levels of rules and procedures that serve as a stable guiding force for the organization.

6.2   Redefining cultures for technology-driven organizations

It is reasonable to derive that based on the result obtained from the analysis of the organization data, technology-based organization cultures cannot be classified as purely mechanistic or organic. Therefore, cultures for these organizations are being classified as composite cultures. A composite culture is being defined as a culture that contains the appropriate level and balance of organic and mechanistic elements and characteristics needed to meet the goals and objectives of the organization. The primary attributes of a composite culture are listed below. For the most part, characteristics of both organic and mechanistic cultures in varying levels were present in the organizations that participated in the study at the time of data collection.

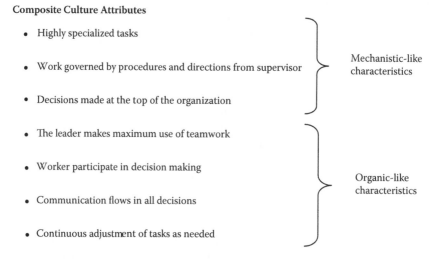

Composite Culture Attributes

- Highly specialized tasks

- Work governed by procedures and directions from supervisor

- Decisions made at the top of the organization

Mechanistic-like characteristics

- The leader makes maximum use of teamwork

- Worker participate in decision making

- Communication flows in all decisions

- Continuous adjustment of tasks as needed

Organic-like characteristics

The organizations that participated in the study culture scores ranged from 3.52 to 5.50. The culture result for each organization is shown in Table 6.1.

The culture data obtained from the study formed the basis for the composite culture theory as shown in Figure 6.1, illustrating where a composite culture will fall on the culture continuum.

(M) = A composite culture consisting of more mechanistic characteristics than organic characteristics.

(O) = A composite culture consisting of more organic characteristics than mechanistic characteristics.

Table 6.1 Culture Means by Organization Data

Organization	Culture mean
G	3.52
A	3.90
D	4.32
J	4.19
B	4.50
F	4.90
C	5.00
L	5.07
M	5.18
K	5.17
H	5.30
E	5.47
I	5.50

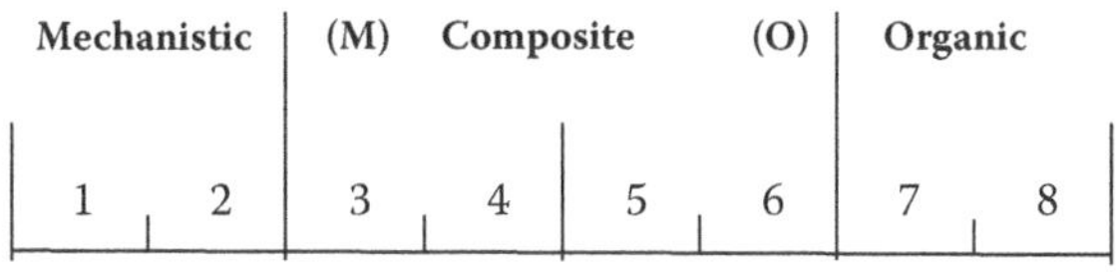

Mechanistic	(M) Composite	(O) Organic
1 2	3 4 5 6	7 8

Figure 6.1 Culture continuum.

Mechanistic characteristics such as having an abundance of procedures and rigid processes in place are necessary in some cases to ensure consistency in operation, product development, and implementation of new processes or programs. Each of the companies that participated in the study found it necessary to have strict control in some areas of the operation to protect the quality of the products and services they provide. Procedures and policies were used to document the practices and the policies that were important to the business to protect the quality of their business line. The companies that participated in the study have been in business for decades and appear to be quite profitable and successful.

The cultures for these organizations fell between the organic and mechanistic regions on the culture continuum. This relationship is shown in Figure 6.2. Observation of the continuum shows that the cultural attributes of the organization were closer to the mechanistic range in most cases, with some of the organizations having cultures more toward the organic region.

Because we now know that trust levels found within an organization depend directly upon the culture operating within that organization, we

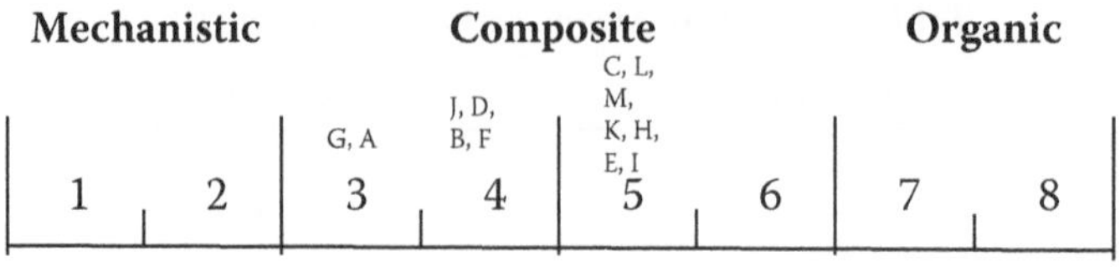

Mechanistic				Composite		Organic	
			J, D,	C, L, M, K, H, E, I			
		G, A	B, F				
1	2	3	4	5	6	7	8

Figure 6.2 Culture continuum—organization placement.

can begin to focus on the type of culture that is optimal for organization growth. The key is designing an organization in a manner to ensure that the appropriate organic and mechanistic characteristics are present and serve the foundation on which the organization functions.

6.3   Culture—the foundation of trust

Human performance in organizations is greatly dependent upon culture. Leaders are instrumental in shaping the culture of an organization. Culture strongly influences the desire to excel or take a passive posture. Some cultures encourage members to be creative and productive whereas others inhibit creativity and productivity. Leaders have the ability to affect their group's culture in such a way that productivity and creativity can be enhanced. Culture is a powerful force that directs the life of its members as individuals and in their relationships with others in the organization. Values supported by culture can form a bond between people more effectively than do formal organizational charts, mission statements, or procedures. A trusting culture facilitates a learning culture, one where people are not afraid to take risks that are believed to be beneficial to the organization. Culture is the foundation on which trust can develop. Equipped with this knowledge, there should be a keen interest in gaining the knowledge and skills necessary to build the foundation so that success can be achieved repeatedly during transactions. The foundation of an organization must have the right balance of elements and attributes that makes it solid and open to trust building.

6.4   Summary

It was established that culture is the foundation for trust to be developed and sustained. Relationship building is also a facilitator of building trust. Oftentimes, the ability or willingness to trust is based on the caliber of the relationship between parties. A trusting environment allows employees to feel empowered to take the appropriate level of risk to be creative and solve problems without the fear of failure. In an environment where people feel free to trust they are more willing to embrace change. A trusting culture also sets the stage for increased productivity. This relationship is shown in the trust paradigm in Figure 6.3.

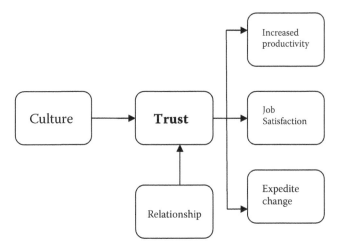

Figure 6.3 The trust paradigm.

The trust paradigm represents an important relationship for management to be aware of and build on when seeking to create an empowered environment for workers to excel in their daily work tasks. When people trust they feel free to open themselves up to new ideas and new ways of doing things.

chapter seven

Study recommendations and conclusions

7.1 Recommendations

Managers should recognize that a culture which facilitates trust is important to their success, because trust can serve as a catalyst to stimulate organizational growth. It is widely accepted that management plays an important role in building trust in organizations. Therefore, they should have a clear understanding of the type of culture operating within their organization, because culture and trust are critical to performance. The knowledge of the type of culture that produces high trust can be important in constructing high-performing organizations.

The results of this study showed that there is a strong correlation between organizational culture and organizational trust. Equipped with this knowledge, managers can introduce elements or attributes consistent with and complementary to their company's business strategy into the culture of their organizations, in order to improve trust. For example, these characteristics may include employing the attributes of trust, increasing the flow of communication across the organization, or utilizing teams to find solutions to problems.

The study's results established that the culture of an organization is affected by the five attributes of trust highlighted by the study. Trust can be improved when management exemplifies the attributes of trust in their daily role as leaders. Management should be cognizant of the impact that the five attributes of trust have on the trust-building process as well as the importance of incorporating these attributes (openness and honesty, concern for employees, reliability, identification, and, competence) in their daily roles to maintain trust.

The trust attributes can also be used as criteria in the selection of new managers. Promoting employees who demonstrate mastery of the trust attributes during interactions can play a significant role in increasing organizational trust. If the right person is selected to move into a management role, these employees are generally viewed by their peers at the onset as trustworthy and competent. Promoting competent managers from within can increase trust among organizational members which can lead to increased performance of the organization as a whole.

Managers can incorporate the trust attributes into the company's leadership training program. Training managers in the attributes of trust can be instrumental in providing them with the skills and knowledge needed to design trusting cultures. A training program of this nature can be an invaluable tool for management because it is widely recognized that engineers and technical professionals lacking the skills needed to function effectively as leaders oftentimes are placed in leadership roles.

7.2   Conclusion

The results of the study demonstrated that the culture of an organization is related to the level of trust found within that organization. The calculated Spearman Correlation Coefficient for the data showed a correlation of 0.79, which indicates that there is a strong relationship between culture and trust. The trust and culture data for each of the organizations was reviewed and showed that the more organic the culture of an organization is perceived, the higher the level of trust. The results specifically showed that employee perceptions of the work culture and trust in their organizations are related.

The attributes of trust most frequently cited by authors and theorists were found to have a relatively strong correlation with the culture of an organization. The attributes of trust most cited in the literature include openness and honesty, concern for employees, competence, reliability, and identification. The correlation results for the individual attributes of trust and culture showed correlations ranging from 0.68 to 0.76, indicating that there is a strong relationship between the trust dimensions and culture. Openness and honesty, the attribute of trust cited in the literature as being of the greatest importance in achieving and maintaining trust in organizations, at 0.76 showed the strongest correlation with culture among the trust attributes. The trust mean result obtained during the study showed that ethnicity, age, job tenure, gender, and job status (management or non-management) does affect trust level in organizations.

A review of the culture results for each organization indicated that the organizations that participating in the study did not exhibit cultures that were entirely mechanistic or organic. The results showed that the organizations exhibited cultures that fell toward the center on the mechanistic–organic continuum. Zamuto and Krakower (1991), in their studies on implementation of advanced manufacturing technologies and the role organizational culture played in implementation, found that cultures combinIng flexibility and controlled values were likely to have structures with both organic and mechanistic characteristics. The results suggest that the cultural environment of most or many technology-based organizations is not entirely mechanistic or organic and that an

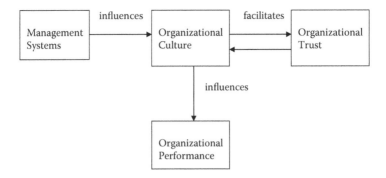

Figure 7.1 Culture and trust roles in organizational performance.

organization can remain competitive in a culture exhibiting both mechanistic and organic characteristics.

The reviewed literature cites many benefits of trust within organizations. Some of the benefits include timely information sharing, increased productivity, commitment, collaboration, and cooperation. The literature supports the premise that designing a culture facilitating trust is one of the most important tasks that managers face. Therefore, it is important for managers to become familiar with the culture of their organizations and understand the relationship between culture and trust in order to build and maintain high-trust organizations. The results of this study, coupled with the current literature, suggest that an organization management system has a direct impact on the culture of that organization, that culture and trust are important to organizational performance, and that these concepts are related. This relationship may be as shown in Figure 7.1.

7.3 Limitation of the study

The following potential limitations of the study may have affected the results obtained and subsequently reported.

The sampled population consisted of a majority of males and Caucasians who were between the ages of 46 and 55. The small representation of other races such as Asians, Hispanics, and American Indians may have slightly affected the overall sample results for these races.

7.4 Areas for future study

The data collected during the study showed a positive relationship between culture and trust. However, additional research would greatly expand the body of knowledge. This expansion will provide knowledge that can be used by engineering managers to design cultures that would increase trust and subsequently improve organizational performance. Additional areas for research are listed below.

- Burns and Stalker (1961) during their research into organic and mechanistic cultures, identified characteristics of each management system. The specific relationship between mechanistic and organic characteristics and the level of trust in organizations should be thoroughly explored. The analysis should seek to identify the impact that each individual characteristic has on building trust.
- The organizations that participated in the study exhibited cultures between mechanistic and organic on the culture continuum. Therefore, the cultures of the participating organizations characteristically can be considered as exhibiting combination cultures (organic and mechanistic characteristics). Consideration should be given to the impact that combination cultures may have on productivity in technology-based organizations.
- The competence attribute of trust deals with the ability of a manager to manage and lead effectively. In order to determine the impact that worker perception of management's competence may play in building and maintaining trust in organizations, an investigation into the relationship between the competence attribute of trust and the characteristics of mechanistic and organic cultural environments should be explored. This investigation should also seek to identify the attributes of a competent manager and how those attributes affect trust.
- An investigation into the relationship between the identification attributes of trust and the characteristics of mechanistic and organic cultural environments may provide more insight into the impact that this attribute has on the various characteristics of mechanistic and organic cultures.
- An investigation into the relationship between the reliability attributes of trust and the characteristics of mechanistic and organic cultural environments should be explored.
- The openness and honesty attribute of trust has been identified by authors and theorists as being the most important attribute of the five identified in building trust in organizations. An investigation into the relationship between the openness and honesty attributes of trust and the characteristics of mechanistic and organic cultural environments should be explored.
- An investigation into the relationship between the concern for the employee attribute of trust and the characteristics or elements of mechanistic and organic cultural environments should be explored.
- In today's highly technical workforce, many projects are performed by teams. The impact of the trust attributes on team culture has not

been explored. An investigation into the role that the attributes play in building high-performing teams should be evaluated.

- The extent of management's role in building and maintaining trust in organic and mechanistic cultures has not been demonstrated and validated through empirical data analysis. An investigation into the relationship or correlation between different management processes such as the Likert management system and the identified attributes of trust can be useful for managers in designing the appropriate work culture that facilitates trust in organizations.
- Retention of technically competent workers is an issue currently faced by technology-based organizations. An investigation into the role that trust plays in the retention of technical workers in combination cultures should be explored.
- An investigation into the role that the attributes of trust play in motivation of employees should be researched. Special attention should be given to the relationship between the Hertzberg (1966) two-factor theory (motivators and hygiene) and trust.
- Relationship building has been viewed as an important element in building trust among people and in organizations. A study to quantify the link between relationships and trust attributes will further add knowledge of the theories of trust and culture. This information can be valuable for practitioners as they lead their organizations.

section three

Practitioner's guide

Part three of this book is written to assist practitioners in their daily roles as leaders to improve the cultural perceptions of their organizations. The guide has a variety of information that can be used to provoke thoughts on effective means to handle the differing situations that are apt to occur daily in an organization.

chapter eight

Tactics for building and maintaining trust

8.1   Introduction

In the wake of organization downsizing resulting from budget reductions and corporate re-engineering, many workers are suspicious of management and do not trust that management has their best interest at heart. It is difficult for managers to gain the trust of workers in environments where change is constant, jobs are being lost, and resources are diminishing. Workers who are not willing to trust management tend to hold on to information that can be helpful in advancing the organization and are not willing to take the risk required to do challenging work. In addition, organizational trust is an important element in determining whether a leader is able to gain access to the knowledge and creative thinking required to solve problems and develop new products and processes.

Building trustworthy cultures starts at the top of the organization with the leadership team. Leadership is one of the most important factors in building conditions of faith and confidence and is generally the most influential in guiding organizational philosophy and behaviors and the work processes within the organizations. Because we know that the culture of an organization is a key feature in determining the level of trust, we can begin to apply that knowledge broadly in terms of designing the appropriate culture of an organization. Once trust is established, people are generally willing to work harder and take on greater risk, which is necessary to develop and create new processes and technologies. This chapter of the book is devoted to providing information to support the appropriate behaviors of leaders that can be used in their attempts to create and maintain trusting cultures. Therefore, various concepts that are deemed important to management and organizational success are introduced and discussed.

There are so many things we do daily as leaders that directly affect the success of an organization of which we may not be fully aware. The net of a leader is widely cast and can have grave impact on the lives of many in the workplace. Managers affect people in organizations in ways such as how work is performed, the way they solve problems, and belief of their work culture. The way managers interact with the worker at all levels must be orchestrated carefully and deliberately. Managers who do not place the appropriate importance on their actions when dealing with people are apt to make huge mistakes that can be costly for an organization. This cost cannot always be measured in dollars and cents, but can be costly in ways that include losing the trust of workers which will affect many aspects of organizational success such as productivity and the willingness of the worker to give all that he or she can to ensure the success of the organization's objectives.

It is commonly known that bad management produces dysfunctional organization, groups, and teams. Dysfunctional organization typically has a negative impact on workers rendering them unwilling or unable to trust. Therefore, it is paramount for leaders to be deliberate in the selection and training of new managers. Oftentimes new managers are selected because of what is referred to as the "just like me syndrome." Managers who are only willing to surround themselves with people who look, think, and act like themselves are hindering and stifling the capability of the organization. They are not viewing the big picture and are most likely operating in survival mode with no long-term vision.

The practitioners' guide is written to provide support to leaders as they work to create trusting cultures and improve organizational performance. The guide contains relevant information on methods and actions that are essential in building and sustaining trust. It also serves as a reminder to managers of the relevance of remaining cognizant of their actions and the impact their actions can have on the worker, teams, and the organization as a whole. The actions of managers are of critical importance because, as we all know, once trust is broken it is difficult or oftentimes impossible to regain. Also included in the guide is a manager's toolkit. The toolkit contains a variety of tools that can be used to determine the level of trust and the type of culture within an organization. The toolkit can also be used to improve trust within organizations and provide managers with options to consider for building long-term trusting relationships in their interactions with subordinates.

8.2   Organizational diversity

The workplace of today is more diversified than in previous years. This diversification of the workplace is expected to continue into the next century and beyond. A diverse workplace consists of workers from different backgrounds including cultural backgrounds, age, race, gender, and experiences, working together to form a more creative, innovative, and productive workplace. In order to tap into the benefits of a diverse culture, attention must be paid to the various cultural backgrounds that are present and the impact that these backgrounds can have on the success of the organization. The variations in social cultures can add to the complexities of building relations that lead to trust in teams and organizations. This change in diversity in the workplace has created challenges for managers in areas such as communication, policy development and implementation, and developing trust. Managers should always consider the role that workplace diversity play in shaping culture, building relationships, and trust within the work environment. Through the literature review and empirical study, we learn that workplace diversity has an important role in shaping cultures and determining the level of trust found within organizations. We explore the impact of diversity and ways managers can use the knowledge gained through the study and literature review in refining the characteristics of the organization in the next sections. Some benefits of a diverse work culture include the following:

- Diversity fosters a more creative and innovative work environment.
- Diversity can serve as a recruitment tool because a diversified workplace signifies that management is accepting of various ideas and concepts.
- Workers with diverse backgrounds bring varying experiences, understanding, and perception to strategy implementation and problem solving.
- Diversity of knowledge and skills can strengthen teams by leveraging the strength of each member sharing different skills and perspectives.

A diversified workplace can be leveraged as an asset for a company. A company can increase its effectiveness when the workforce is representative of the company's customer base. Forward-thinking companies are skilled in how to leverage a diverse workforce. For example, team members with knowledge of multiple languages when conducting business internationally can be an asset in bridging the language and cultural gaps that may prevent successful business interactions and negotiations. Also, a culturally diverse team or organization will have unique ideas about approaches to marketing, products, and services that will appeal to the people of their respective cultures. Businesses are better able to breed the type of creative and innovative solutions when they bring together different backgrounds, skills, and experiences. These businesses typically realize a significant increase in worker job performance and retention of critical skills. Embracing diversity in organizations has many positive impacts on an organization including increased team learning and increased trust in management and the organization.

8.2.1   Gender intelligence

Psychologists and sociologists have pointed out routinely that men and women are different in the way they think, act, and process information.

If one were to observe men and women in verbal and nonverbal communication the differences may become evident. Therefore, it is not a surprise that the study validated that men and woman within the same organization displayed a different perception of the culture and displayed different levels of trust. It is important for managers to understand that males and females will in many cases have different perceptions of their work culture and will show differences in the level of trust in management and their organization. This difference in perception can also be linked to and seen in the behaviors and practices of the members of an organization.

The results of the study presented in Section II of this book showed that trust levels for women working within the same organizations were not always consistent with those of their male counterparts. Women continue to enter the workplace in record numbers from various socioeconomic backgrounds and are becoming an integral part of the success of organizations. Whether females are able and willing to integrate and develop trust in their organization can be an important factor in retention of critical skills. The results of the literature review also showed that women and men communicate differently. This difference in trust between males and females can be attributed to various attributes of the organization culture such as policies and procedures, practices, behaviors, relationships, and communication channels. For example, in a male-dominated culture or in a culture where 90% of the leadership team consists of males, females may feel less appreciated and less valued if opportunities for advancement are not equally afforded to them. As a result, women may be more skeptical and may require more information or data before providing buy-in for key concepts. Listed below are some actions managers can take to improve perception among females and develop trust within the organization and in the management team.

- Ensure that females are included in the decision-making process.
- Provide career opportunities for females equal to those offered to their male counterparts.
- Communicate openly and honestly with the appropriate amount and level of detail.
- Provide equivalent salaries for females to those offered to their male counterparts.
- Refrain from making gender-based jokes and gestures that target females.
- Provide the same caliber of respect to females as shown to their male counterparts.
- Do not institute policies or practices that are geared toward one gender. For example, females with a family may not be able to

work late. Therefore, a practice of scheduling late meetings may not be appropriate for a work–life balance and can provide in some cases an unfair disadvantage for females.

8.2.2   Ethnicity and trust

A discussion on diversity in organizations is not complete without addressing the role that ethnicity plays in forming the work environment. The study revealed that there was a difference in the level of trust among the ethnic groups represented by the study. As a reminder, the result of the study is listed in Table 8.1. This variance in trust level can be in part a result of the social cultures these groups bring to the workplace, the actions of the leadership team, policies, procedures, behaviors, and norms of the work environment.

The U.S. work population for workers from countries outside the United States is constantly increasing. These workers bring to the work environment various attributes of the social culture of their homeland. In order to formulate trusting relationships among the diversities of cultures, managers need to understand the social cultures that are part of the organization. Below are some actions that managers can take to begin addressing the consideration for ethnicity in their work culture.

- Understand the diversities of social culture of the workforce.
- Provide equal opportunity for pay, promotion, training, and rewarding assignments.
- Include ethnicity representation in the policy setting and decision-making processes.
- Learn important critical attributes and traditions of each social culture.
- Celebrate and embrace the various social cultures.
- Respect the various social cultures represented in the organization.
- Avoid putting policies in place that could be viewed as culturally offensive.

Table 8.1 Ethnic Group Analysis

Ethnic group	Trust level
Asian	3.80
American Indian	3.74
African American	3.65
Caucasian	3.53
Hispanic	3.42

8.3   Organizational alignment and policies

One may ask what do organizational alignment and polices have to do with creating trusting cultures. The establishment of policies and procedures can help an organization demonstrate its commitment for requirements such as diversity, ethics, and regulatory compliance. Consistent implementations of these policies, procedures, and regulations directly help form and shape culture. The establishment of human resource (HR) policies outlines obligations and the standard of behavior for the management team and each employee. Organizational alignment can create complexities in communication channels, the policies that dictate the practices, and the way they are carried out. Some of the most important aspects of these policies are the HR policies, implementation of these policies, and the practices that are not governed by documentation. Although HR policies are primarily developed by the human resources staff, these policies are endorsed and are implemented by managers. As we have established, the actions of the management team are an important element in building cultures and developing trust. Therefore, managers need to endorse and implement policies that support building trusting relationships and the needs of the organization as a whole. A firm's policy can be very effective in supporting and building culture if

- The leadership team and the workers buy in to the policy or procedure.
- The policies have a purpose and make sense.
- The policies are implemented consistently.

Managers are primarily responsible for ensuring policies are aligned and implemented fairly and consistently. Inconsistent policies and procedure implementation can lead to high employee turnover, lower productivity, and workers not willing to trust their leadership team. Procedures and policies should be strategically developed and implemented because they can have an impact on the worker and the culture of the organization. There needs to be a balance between policies and procedures and allowing employees enough autonomy to be able to create and invent new processes and technologies.

8.4   Dealing with perceptions

Is it perception or is it reality? This question is commonly asked when dealing with various issues in life as well as in an organizational setting. Many times managers are quick to dismiss situations or issues that they believe to be based on perceptions. They are of the opinion that time spent dealing with or addressing these issues is a waste of valuable time. This is a huge mistake! Trust in organizations can be developed and sustained based on individual perceptions of their work culture. In such cases, perception can become the reality for a manager. In the workplace, perception can quickly appear to be accurate regardless of whether it's fictitious or accurate. Perceptions in the work environment can have an impact on culture, trust, retention of employees, and whether a manager is believed to be competent. These perceptions are formed based in part on observed actions of workers and the leadership team. Perceptions can be managed if addressed expeditiously and appropriately.

Management styles can have a lasting impact on workplace perception. For example, a manager can be perceived as uncaring if his or her leadership style is perceived to be hands-off, and a detail-oriented manager may be perceived as not trusting employees to do their jobs. The actions of managers can serve as a conduit to improve positive perceptions and counteract some negative workplace perceptions. The following actions can assist in dealing with perceptions in the work environment:

- Address perceptions as you would address known issues.
- Communicate openly, honestly, and frequently.
- Model the behavior expected of others.
- Demonstrate competence in leadership.
- Refrain from showing favoritism.
- Treat everyone fairly.
- Respond to questions and issues promptly.

8.5   Role of trust attributes in building a culture of trust

During the two empirical studies and the literature review it was validated that the five trust attributes are very important in building trust in organizations. In addition, the relationship attribute has been added inasmuch as it also has important applicability in the trust-building process. These attributes are listed and briefly defined in Table 8.2. Building and

Table 8.2 Trust Attribute Definitions

Trust attributes	Meaning
Competence	Effectiveness of workers and leaders
Openness and honesty	Amount, accuracy, and sincerity of information communicated
Concern for employees	Demonstration of empathy, tolerance, and safety
Reliability	Consistent and dependable actions
Identification	Sharing common goals, values, and beliefs
Relationship	Ability to build and sustain respectful relationships

Table 8.3 Trust Attribute Correlation with Culture

Trust attribute	Correlation
Openness and honesty	0.76
Competence	0.70
Concern for employees	0.73
Identification	0.68
Reliability	0.68

nurturing relationships is an integral part of the trust-building process that is often overlooked by leaders. Years of leadership experience has taught many practitioners that trust can develop among people depending on the type and nature of their relationships. The ability to build trusting relationships can also be a factor in determining whether subordinates are willing to follow their leader in times of uncertainty.

Each of the trust attributes serves as an integral part of the overall process of building and sustaining trust. In the next sections we explore each attribute and highlight how it can be used by management in building and retaining trust in organizations. Also demonstrated through the study, the five trust attributes highlighted correlated strongly with culture as shown in Table 8.3. Although the sixth attribute, relationship, was not included in the study, it is recognized as a known important attribute in building trust. Knowing the connection between the attributes of trust and culture, managers can incorporate the trust attributes in their strategy to build and maintain a trusting culture. A detailed discussion of each of these attributes (openness and honesty, reliability, identification, concern for employees, competence, and relationship) follows.

8.5.1 Openness and honesty

We begin our discussion of the attribute of trust with the openness and honesty attribute. This attribute is cited most frequently by authors and theorists as being of the most importance in the trust-building process. In addition, this attribute correlated highest with culture during the empirical study. Good relationships can develop when people are open and honest with each other. Openness and honesty are important characteristics and qualities for building and maintaining any relationship. Together they form the foundation for trust. Openness implies giving another person access to one's self, thereby allowing oneself to become vulnerable during communication. Honesty is providing truthful information during communication when asked and openness is offering the truth without being asked. Honesty in the workplace creates an environment for creativity and productivity to flourish. Openness and honesty can be demonstrated by the following:

- Be completely honest in communication.
- Communicate clearly and concisely.
- Provide clear, concise, and honest feedback.
- Communicate information in totality.
- Show a willingness to share information.
- Be receptive to new ideas and thoughts.
- Remain accessible to your staff.

In organizations where managers are viewed as honest, workers are willing to go the extra mile to ensure success of the organization. Open and honest communication in the workplace fosters innovation, collaboration, team building, and transparency. Management is an important variable in fostering open and honest communication in the workplace. Therefore, they themselves must "walk the talk."

8.5.2 Competence

Managers have little to no credibility if they do not have a high level of leadership competence. This does not mean that the managers must be experts in the technical aspects of the area they have been charged to lead. Leadership competence refers to the ability of the manager to lead the organization effectively. This ability to lead encompasses the leader's ability to make good sound decisions as well as the ability to engage in effective communication internally and externally. Some characteristics of a competent leader are listed below.

- A competent leader sets clear expectations.
- A competent leader is humble.
- A competent leader takes calculated risks that are beneficial for the organization.
- A competent leader listens with an ear to hear.
- A competent leader is positive in actions and interactions.
- A competent leader will support what is right rather than the popular decision.
- A competent leader routinely praises his people for doing well.
- A competent leader is not afraid to say thank you.
- A competent leader can admit his mistakes.
- A competent leader has a vision.
- A competent leader can handle negative and uncomfortable communication and situations.
- A competent leader holds the workers accountable for the tasks with which they are charged.

- A competent leader is interested in the well-being of others.
- A competent leader is approachable.
- A competent leader is honest.
- A competent leader is trustworthy.
- A competent leader looks for opportunities to promote and help others grow.

Oftentimes leaders promote people who are technical experts in their fields without assessing their leadership competence. Many technical experts lack the ability and fortitude to function as competent managers due to lack of training and experience. However, these competent and bright technical professionals end up being promoted to supervisory or management positions while not being adequately equipped for success. Because these professionals oftentimes lack the skill to lead and manage they are not viewed as competent managers by the people they have been charged to lead. Competent managers should exhibit the following characteristics:

- Demonstrate leadership and management skills.
- Lead with confidence and assurance.
- Decisive in decision making.

8.5.3   Concern for employees

Managers who demonstrate concern for employees are well on their way to developing a strong relationship that can lead to an environment where trust flourishes. Employees want to know that they are important to management and the organization as a whole. Managers who demonstrate that they have the best interest of the workers at heart are able to tap into knowledge, creativity, and their interest in ensuring that the goals and objectives are met. Some ways that managers can demonstrate to employees that they are concerned about their health, safety, and well-being are presented below.

- Provide meaningful and rewarding opportunities.
- Ensure appropriate and fair compensation.
- Provide adequate benefit programs.
- Support work–life balance.
- Listen attentively to ideas and concerns.
- Keep workers informed.
- Provide a safe and clean workplace.

8.5.4   Identification

Trust is facilitated when individuals share an association with or assumption of the qualities, characteristics, or views of another person or group.

For the purpose of this discussion identification is being defined in terms of organizational solidarity. Solidarity refers to the integration shown by the organization and its members and the ties that bind them together which form the basis of the organization's culture. These ties often include the vision and the values of the team. When employees feel that they can identify with the management team and the vision of the organization it is easier for them to feel connected to the leader and the organization. Identification can be achieved through the following:

- Engage in appropriate casual nonbusiness discussion with the employee.
- Show an interest in the project or task being performed.
- Demonstrate buy-in for the goal of the organization.
- Demonstrate a sincere buy-in for the vision and mission of the organization.
- Follow all policies and procedures.

8.5.5 Reliability

Reliability in leadership means not only showing up for work but being prepared to lead and inspire. In addition it also means being someone who can be counted on based on words and actions. Leaders who are viewed as unreliable will never be seen as leaders who can be depended upon and trusted. As demonstrated by the study and the literature review, reliability is an important element in building trusting cultures. The study results reiterated the importance of reliability in building trust in organizations. As a reminder, the results showed that the reliability attribute appeared to be one of the most important attributes for most of the technology-based organizations that participated in the study. Due to their erratic and unpredictable behaviors, too many managers contribute to the erosion of worker trust. Dealing with an erratic and unpredictable leader can be stressful for most workers. Consistency is a key element in being viewed as a good leader. A consistent leader sets the tone and makes decision making and problem solving by subordinates easier. A leader can demonstrate reliability through the following actions:

- Keep the promises you make.
- Be consistent; employees will come to know and understand what is expected and the value systems that guide decisions.
- Be predictable.

8.5.6 Relationship

Another important aspect of building trust includes the ability to build lasting and effective relationships. Building trusting relationships that last over time takes practice, time, and patience. People are different when it comes to their need for close personal relations with others. Most people tend to proceed with caution when developing or embracing relationships in the workplace. An organization can be described as a network of relationships that requires all parties to contribute in order to achieve a common objective. This network of relationships is the basis of forming a culture where trust can develop and grow. The culture of an organization can help define the acceptable level of personal relationship within the organization. An organization's culture based on cohesive relationships, trust, and respect fosters attachment among individuals. Successful managers strive to develop a relationship based on trust with colleagues, employees, and customers. Leaders who have formed trusting relationships with employees can ease or in some cases eliminate some of the frustration associated with a rapidly changing work environment. Even when change is not completely embraced by all, employees will trust and follow a leader who has invested the time in building a relationship with them and is perceived as being honest, approachable, and trustworthy. Trusting relationships generally motivate and bring out the best in people. Motivation resulting from trust can stimulate innovative thinking and creativity that leads to increased organizational improvements and increased productivity. The ability of a leader to enter an organization and develop a relationship that leads to trust is critical to move the organization forward successfully. Some managers believe that respect and followership result from the nature of their leadership position. This is not necessarily an accurate assumption because people generally reserve the option to trust and follow based on their relationship with and perception of the leader. When a leader enters an organization, many or most members of the organization take on a reserve "wait and see" posture. Meaning, they reserve their opinion until the manager has proven himself or herself in their view. Trusting relationships are built over time. Ways to build trusting relationships include the following:

- When communicating, separate facts from feelings.
- Create an environment or atmosphere where you can trust and share openly without being afraid.
- Demonstrate respect.
- Listen actively and openly.
- Promptly address disagreements.

8.6   Selection and promotion of new managers

Selecting new managers is another important task leaders perform in planning for longevity and success. Selecting the right person to become a future leader is instrumental in ensuring the viability and survivability of an organization. Yet this process is not oftentimes conducted with the appropriate amount of care and strategy. Some leaders select managers who are "just like them." They select managers who act like them, think like them, and approach problems like they do. This selection process produces managers who may not be completely supportive of diversity and lack the ability to plan strategically for the future. These managers tend to manage from day to day with no vision or strategy for the future of the organization. Therefore, because they may not be prepared for the challenge, these managers are incapable of facilitating the type of culture needed to develop and maintain trust. Newly appointed managers do not automatically become excellent managers by the act of their appointment or promotion. These managers need to be provided the tools necessary to be successful. A good training and mentoring program is the beginning of developing these managers into great leaders of the future. Elements that should be considered in designing a good training program for future leaders are addressed in Section 8.6.1.

8.6.1 Leadership training

A good leadership training program is considered an asset for companies as an increasing number of technical professionals move into leadership positions. These technical professionals do not necessarily possess the skills to lead effectively at the time of promotion. Many companies currently are offering leadership training for new supervisors and managers as well as for long-term experienced supervisors and managers. However, oftentimes this training focuses on general leadership skills such as communication, strategy development, scheduling, planning, handling a budget, and so on. Although training in these areas is critical, it is does not provide all of the knowledge needed to lead people and an organization successfully. These courses do not focus on training managers on the methods that can be used to build trusting cultures. Nor do they address the importance of lasting relationships and ways new managers can build relationships that lead to trust between them and their colleagues, customers, and subordinates. In addition:

- These courses do not address the importance of culture and the ways cultures are formed and managed.
- These courses do not address the importance of trust and ways trust can be developed and maintained.
- These courses do not address in detail the importance of relationships that will foster and sustain trust.

- These courses do not address the importance of managers' actions and the roles a manager plays in creating environments that can erode trust, reduce productivity, and critical thinking necessary for growth.
- These courses do not address ways that management can deal effectively with perceptions in the workplace.

The result of the study established that a trusting organization is not possible without the appropriate culture at the root of the organization. The result of the study also showed that as the culture of an organization moved toward becoming organic on the continuum, the higher was the level of trust present within the organization. Simply put, individual perception of the organization's culture directly affects the level of trust in management and the organization. Therefore, managers should be trained in the aspects of culture that facilitate trust and the ways trust can be developed and maintained. Some of the attributes that should be included in the training provided to managers and supervisors are the following:

- Building trusting relationships
- How to build and sustain trusting cultures
- How to demonstrate the trust attributes in managers' daily role as leaders
- Ways to deal with cultural perceptions

8.7 Building trusting cultures

A very important aspect of culture is the connection that exists among leadership, the worker, and the culture of their respective organizations. Culture determines behaviors that bond, motivate, and set the stage for trust to be cultivated and flourish. In order to gain an understanding of the importance of creating trusting cultures in an organizational setting, it is necessary to understand the role leadership plays in developing or forming and maintaining these cultures. Building trust within an organization starts at the top of the organization. Leadership is the most important factor in facilitating and maintaining credence and confidence.

One of the most difficult aspects of building a competitive team and organization is managing the culture because values, principles, and practices are not easily recognized and managed. The way in which policies are implemented and the behavior of leaders can provide an indication of whether the culture of an organization is perceived to be based on trust or fear. Fear-based cultures produce members who are not willing to take chances and therefore creativity is stifled for fear of failure and the retribution that may occur in the event of that failure. Leaders must invest the time and energy in building a culture that can serve as a springboard to ensuring organizational success. We all agree that building cultures that encompass elements such as trust, productivity, creativity, employee involvement, and so on is not easy, however, they are necessary for competing in a changing global environment.

8.8   Perils of mistrust

In many environments it seems to be easier for mistrust to develop. Mistrust has a serious negative impact on the attitudes of people and their willingness to buy into the vision and values of the organization. Generally mistrust develops in an organizational setting as a result of the actions of managers, the policies that are developed, and the way those policies are implemented. Many times policies are not implemented or are perceived as not being implemented consistently across the board. Some common actions exhibited by leaders that can lead to mistrust in the work environment are the following:

- Display of inappropriate behavior or conduct such as abrasive or abusive interactions with colleagues and subordinates
- Disregard for the organization's policies and procedures
- Lack of follow-through with what has been communicated
- Inaccurate or incomplete communication
- Display of ambiguous, unclear, unpredictable, and erratic behavior by management
- Employees' perceptions about the leadership team

These behaviors can hinder or delay the development of trust in relationships and in organizations and can be damaging to the company's bottom line. There are actions and signs that signal the presence of mistrust. Below are some warning signs that distrust is at the core of the culture of an organization.

- High employee turnover
- Low initiative
- Low morale
- Defensive attitude and actions
- Fear among the employees
- Enormous approval process with unnecessary checks and balance

8.9   Behaviors that facilitate trust

Working with people and solving problems continue to be critical roles for managers. A worker's belief about his or her leader's character, abilities, and intention has an important impact on trust. A leader must exhibit

Table 8.4 Trust Behaviors and Outcome

Behavior	Expected outcome
Honesty.	Promote respect and trust.
Honor agreements.	Promote credibility and trust.
Open communication.	Provide clarity to expectations and facilitates trust.
Admit mistakes when they are made.	Promote respect and a sense of humility and humanity.
Provide honest and constructive feedback.	Facilitate trust.
Be willing to give and able to accept constructive feedback.	Demonstrate humility.
Maintain confidentiality.	Facilitate trust and open communication.
Delegate.	Demonstrate trust and respect for people's knowledge, skills, and ability; this eliminates the tendency to micromanage.
Involve others and seek their input.	Promote ownership and pride in work.
Be consistent and reliable.	Build confidence in leadership ability and behavior.

trustworthy behavior in order to gain and retain trust. Trust within an organization is based specifically on the caliber of leadership within the organization. When building trust, it is necessary to pay attention to every aspect of an organization's structure such as management policies and practices, values, and expectations of its members and especially the behavior of its leaders. There are many behaviors exhibited by managers that can help improve trust. Some of these behaviors and expected outcomes are listed in Table 8.4.

8.10   *Interpersonal communication*

There is no profession that does not require communication and interaction with others. Interpersonal communication is complicated due to factors such as the diversity of cultures in the work environment. People from varying cultures communicate differently, therefore not being knowledgeable of the various cultural traditions can limit the ability to engage in effective communication. Communicating with others is an essential skill in business dealings and building effective relationships. Interpersonal communication skills are an important attribute in developing relationships within organizations. This type of communication involves an exchange of information and feelings through verbal and nonverbal means. Interpersonal communication is not only about what is communicated or the language that is used; it is also about the unspoken words represented by body language and facial expressions. A good understanding of interpersonal communication is an essential part of developing lasting relationships and trust during interactions. The following should be considered to assist with improving communicating:

- Be aware of gestures during communication.
- Control tone of voice.
- Listen actively.
- Be cognizant of facial expressions, eye contact, and eye movement.
- Remain cognizant of body language.
- Understand the culture of the person with whom you are communicating.

8.11   Soft side of management

The "soft" side of management is the caring relationship and leadership-friendly part of a manager based on emotions. The firmer strategic side of management is the role of management that makes plans, sets up structures, and firmly monitors performance. The real challenge for most leaders is being able to maneuver between displaying the firm side versus the soft side. Overemphasizing the firm side of management can lead to fear and mistrust among workers. Being an effective leader does not come easily for many because often managers are unable, unwilling, or uncomfortable with revealing their softer side. The softer side of management involves acting upon intangibles that can sometimes be seen based on actions or reactions. On the other hand some intangibles such as empathy, intuition, and the ability to access the needs of others are not visible. Effective managers know when and how to display the appropriate level of vulnerability when communicating. Building relationships that lead to trust oftentimes means that managers are willing to display the softer side of their personalities, letting others see their humanity so that a connection can be made and trust can develop. Showing one's humanity does not mean that weakness exists. On the contrary, it demonstrates that the manager is confident and secure in his or her knowledge, abilities, and competence as a leader.

8.12   Leader's role

Management effectiveness depends on the ability to gain the trust of subordinates and colleagues. When trust is relatively high, employees are more committed to authority and are willing to do whatever it takes to help the organization succeed. An employee's attitude toward change and ability to maneuver through the change management process is affected by his or her trust in management. The more an employee trusts management, the easier the change management process becomes. Some leaders make decisions that are based solely on their connection and relationship with others. These decisions will unintentionally have an adverse effect on building trust with the majority of people in the organization. Another scenario to consider is that there are times when some leaders make decisions based on their needs and what is easiest for them. Such actions should be avoided because of the damaging effect it can have on the workers.

There are many benefits achieved when employees trust their leaders. These benefits affect the entire organization on various levels. Some of the obvious benefits of having trusted leaders in an organization are listed below.

- Leaders who are trusted empower workers to think and make decisions.
- Leaders who are trusted facilitate productivity because workers are inspired to create and produce without fear of failure.
- Leaders who are trusted encourage focus on the task to be accomplished.
- Leaders who are trusted ignite passion and excitement for the task at hand.
- Leaders who are trusted foster innovation and new ways of solving problems to help the company move forward and successfully compete in their business.
- Leaders who are trusted can be instrumental in facilitating an employee's decision to remain with an organization.

Trusted leadership is contagious because when trust is present within an organization your customers and business associates know it and they will respond accordingly. Therefore, they are more willing to trust the organization. This level of trust among the customers leads to repeat work and can open up additional external untapped opportunities that arise through word-of-mouth recommendations from current clients. Listed below are actions that leaders can take to build and maintain trust.

- Demonstrate concern for employees.
- Be predictable.
- Develop trustworthy relationships.
- Ensure your actions and words match.
- Keep confidences.
- Do not gossip.
- Treat all employees fairly.
- Communicate often and completely.
- Confront issues in a timely fashion.

- Display competent leadership skills.
- Act with integrity.
- Keep commitments.
- Display competence in management skills.
- Listen attentively and with respect.
- Exhibit empathy and sensitivity to staff needs when appropriate.
- Lead with humility.
- Apologize when you make a mistake.
- Fulfill promises.

8.12.1   Performance appraisals

Conducting performance reviews is an integral part of a manager's role. Most employees dislike the process and consider it to be cumbersome, a waste of time, and not necessarily an accurate account of their performance. If not performed appropriately, performance reviews can drastically reduce employee motivation to buy into the vision of the organization and limit productivity, as well as inhibit trust in management and the organization. A performance appraisal (PA) is a systematic process that is used periodically to evaluate an employee's job performance. The review is evaluated against pre-established goals and objectives. A documented PA is typically conducted annually. However, it is a good practice to conduct a mid-year review to provide feedback to employees on their performance for the first six months of the performance period.

PA data are generally conducted using two primary methods; objective data and judgment of the responsible manager. In order to have a meaningful PA experience, good preparation is needed from both the manager and the employee. The following should be incorporated into the PA review process to bring ease and meaning to the process.

- Ensure that enough time is allotted for the PA discussion.
- Meet in a private place.
- Review all documents from the previous PA cycle.
- Review and have on hand the PA discussion documents and notes accumulated during the year (include awards, achievements, and feedback from customers and co-workers).
- Discuss employee success since the last PA review.
- Discuss obstacles that the employee may have encountered since the previous PA.
- Discuss all training needed and completed to ensure employee success.
- Discuss ways you as the manager can assist the employee in achieving his or her goals.

- Listen attentively.
- Provide open and honest feedback.

8.12.2   *Reward and recognition*

One of the most expeditious and effective ways to express appreciation is often one of the most simplest and overlooked: saying *"thank you."* Thank you is appropriate and effective even when most of the duties performed are viewed as a normal part of the job. Hearing *"thank you"* represents a spontaneous and timely yet effective means of expressing gratitude in a way that can mean a lot to anyone. Expressing gratitude should be done often and with sincerity, and can be done privately or publicly in front of co-workers. When expressing gratitude ensure you mention the task, project, or behavior you are recognizing. There are various ways to recognize groups, such as picnics, luncheons, breakfast celebrations, plaques, certificates, and financial incentives. A good reward and recognition program can be the catalyst to helping employees recognize how much their contribution to the success of the organization is valued and appreciated. Employees who feel valued are more open to supporting and trusting management and the organization.

8.12.3   *Succession planning*

The succession planning process is an important process to ensure that an organization retains the talent and skilled workers need to carry on the business of the organization. A leader's ability to plan strategically is another measuring stick that can be used to determine if a leader is to be viewed as being competent. Succession planning refers to the process or strategy used to identify and develop potential successors for key positions within a company. Key positions may include leadership positions or highly skilled technical positions that are difficult to replace. Oftentimes key positions may require specialized training, license, or experience level that is difficult to replicate. An important aspect of succession planning is to create a suitable match between the company's future needs and the desired career path of individual employees. A well-developed succession planning process can serve as a means to increase the retention of high-performing workers because through succession planning the value of the worker is being demonstrated by the company's investment in employee development. This is another process that can serve to reduce motivation and productivity as well as decrease trust in management and the organization if not properly carried out. Therefore, it is incumbent upon managers to approach this process carefully and communicate appropriately. This is yet another opportunity for the "just like me" syndrome to rear its ugly head. Careful consideration must be given to the selected successor.

8.12.4 Employee engagement

When many people think of employee engagement, they immediately think about granting the employee the ability to participate in all business decisions. Employee engagement refers to the active involvement of employees in shaping the culture of the organization and participating in the success of the organization. There are some decisions that are necessary to be made by senior management on behalf of the employees. However, employees should be actively engaged in some decisions that affect them and the organization. When referring to employee involvement, focus is being extended to creating an environment in which people have an impact on decisions and actions that affect their jobs as well as being involved in some decisions in the business dealings of the organization they support. Employee engagement is a leadership philosophy involving practices that demonstrate people are valued and are invited to contribute to continuous improvement and the ongoing success of an organization. Employee involvement has these benefits for an organization.

- Facilitates employee growth
- Empowers employees to participate in the business
- Increased trust in management
- Increased employee retention
- Increased productivity

8.13 Leadership styles

Leadership style plays an important role in determining whether a manager is able to build relationships and gain worker trust. We know that workers will follow a leader they trust even when times are uncertain and difficult. There are many leadership styles a leader can assume during interactions with subordinates and colleagues. At times it is necessary to change styles based on the individual or team that is being led and the accompanying circumstances. There is not a "one size fits all" style when it comes to leading people. There are times when leaders must adapt their styles to the situation at hand. Effective leaders know when they need to change their style in dealing with the many challenges they face. We will explore three leadership styles and their impact on building trusting cultures in an organization setting. The three leadership styles discussed in the upcoming sections are situational, servant, and the micro leader. These leadership styles were selected for discussion because they can have the greatest impact on building or eroding trust in organizations.

8.13.1   Situational leadership

Situational leadership entails the adjustment of leadership styles by managers to fit the development level of the followers they are is trying to influence. With situational leadership, it is incumbent on the leader to change his or her style, as opposed to the follower adapting to the leader's style. Situational leadership entails four predominant leadership styles. These styles are discussed briefly below.

- *Telling:* Leaders tell their subordinates what to do and how to complete a task. During this stage the leader has an opportunity to build relationships with subordinates and create an atmosphere for trust to develop.
- *Selling:* Leaders still provide information and direction, however, there is more communication with the follower. Leaders work to get followers on board. This stage presents the opportunity for managers to begin demonstrating leadership competence and developing trust.
- *Participating*: The leader works with the team and shares decision-making responsibilities. The leader focuses more on the relationship with the worker.
- *Delegating:* Leaders pass on many of the responsibilities to the subordinates. The leader assumes the role of monitoring progress and has less involvement in making decisions. During this stage the leader has an opportunity to demonstrate trust in the worker which can solidify the trust developed during previous interactions. The leader has an opportunity to demonstrate trust in subordinates that has developed and been nurtured during the first three stages.

The situational leadership style is optimal for building trust through use of the telling, selling, and participating styles and reinforcing trust through use of the delegating leadership style.

8.13.2   Servant leadership

A servant leader demonstrates value for everyone's contribution to the decision-making and problem-solving processes and often seeks out the opinion of others. Servant leaders are completely devoted to serving the needs of the organization and the people they lead. A recognized strength of this leadership style is that it steers a manager away from taking on a self-serving attitude. Some characteristics of a servant leader are listed below.

- Values diversity in decision making
- Skilled in the development of others; deeply committed to professional and personal growth of others

- Cultivates trusting cultures
- Helps people with issues
- Encourages others to excel and succeed
- Listens attentively; committed to listening to others; listens receptively to what is being communicated
- Thinks strategically and long term
- Acts with humility
- Thinks of others as opposed to himself
- Sells ideas as opposed to telling; relies on his or her persuasive ability to build consensus within groups

The servant leader style, practices, and philosophies place a leader in a good position to build and maintain trusting relationships because the primary focus is on others as opposed to him or herself. The servant leader shares information and power, puts the needs of others first, and focuses on helping people develop and perform to their highest capacity.

8.13.3   The micro leader

A micro leader is a leadership style where a manager closely observes and controls the work of subordinates to the point of stifling organizational growth. Most people who have been in the workforce for any length of time have been exposed to a micro leader. These leaders are often recognized quickly by the people they manage. Managers with this style of leadership can shake your confidence in your own abilities and create a co-dependent workforce. Once worker confidence is shaken, the worker generally becomes timid and afraid to move forward without guidance. These leaders are known by their peers and subordinates as "power hungry" with the need to control everything and everyone. The micro leaders tend to have a problem trusting people and people tend to have an issue with trusting micro leaders. Signs and characteristics of a micro leader are listed below.

- Resists or is unable to delegate
- Corrects tiny details as opposed to looking at the big picture
- Discourages others from making decisions without consulting him or her
- Overly involves him or herself in overseeing the projects of others
- Places the worker under a microscope
- Requires subordinates to always consult with him or her before making a decision
- Makes all decisions without considering input from others
- Has all of the answers and is never wrong
- Does not allow subordinates to have autonomy to complete their work

Competent and effective managers set up those around them to succeed. Micro leaders, on the other hand, prevent employees from making decisions and taking responsibility for their decisions. People tend to grow when they are allowed to make decisions and deal with the consequences of those decisions. Micro leaders restrict the ability of people to grow and limit what the organization can achieve inasmuch as every decision must be funneled through the manager. In addition to creating stress and discontent among employees, the micro leader's style inhibits employee development, effective relationship building, and building trust.

chapter nine

Manager's toolkit

9.1   Introduction

The manager's toolkit is designed to assist managers in assessing culture, improving relationships within their organization, and building a culture that serves as the foundation for trust to develop and grow. The kit contains a variety of resources that will be helpful in shaping culture and improving trust. The toolkit includes the following:

- *Case studies:* Included are five case studies that simulate the type of interactions and activities that can happen in an organizational setting between managers and subordinates. Analysis of these case studies should provide insight into how managers could properly handle similar situations that may occur in their respective organizations.
- *Assessment guide:* The assessment guide provides instructions on how to assess culture and trust. An assessment should be performed in order to understand the cultural health of an organization. The assessment guide is structured to provide basic knowledge of ways to perform an effective cultural evaluation. The information gained through the assessment should be used to improve the culture of the organizations.
- *Survey questionnaires:* Used to provide an indication of various attributes of the organization. The information obtained through the use of the questionnaires can provide important information on the environment in which work is performed as well as the perceptions of the workers.

9.2   Guidelines for assessing organizational culture and trust

A self-assessment of organizational culture and trust is an effective means to examine the attitudes, perceptions, practices, and policies that are in place to support the culture of an organization. The knowledge gained from an assessment is a key component of the continuous improvement process. Organizational culture and trust should be assessed in three parts to gain the maximum information needed to evaluate and attain useful information that will assist in modifying the culture if necessary. The first part should consist of a review of culture-supporting documents. The second part should be through employee surveys followed by focus group surveys and individual interviews. It is important that prior to embarking upon the assessment, employees are made aware of the pending assessment and its importance to management and the organization. This is an opportunity for management to be open and honest with the workers and request their open and honest feedback when completing the survey. The following detail should be provided to the workers prior to beginning the assessment:

- Why the survey is being conducted
- How the survey will be administered
- How the focus group discussions will be conducted
- How people will be selected to participate in the interview portion of the assessment
- What to expect during the focus group survey
- How the result will be used
- When the results will be analyzed and published
- How the results will be communicated

9.2.1   Document review

The first step in assessing culture is to review the documents that detail the processes, practices, and behaviors expected by the organization members. These documents should be reviewed to ensure that the appropriate expected behaviors are clearly defined. Policies and procedures documenting the roles, responsibilities, and accountabilities for organization members should also be included in the review.

9.2.2   Selecting a survey instrument

There are many survey instruments that are available to measure culture and trust. Survey selection and development should be based on the type and amount of feedback you would like to get from the survey. The survey instrument should contain the detailed questions that are designed to capture the perceptions of the individuals completing the survey. These questions are specifically designed to capture specific types of information that will be helpful in improving organizational performance. Although there are several survey instruments available to measure culture and trust, these instruments measure various elements or attributes of culture or trust. Therefore, it is necessary to select the instrument carefully that will complement the scope of the assessment. When selecting a predeveloped survey the following should be considered:

- Cost associated with using the survey.
- Does the survey ask the right questions?
- Does the survey contain questions that are clear?
- The method used to analyze the data.
- The method used to administer the survey (electronic or on paper).
- Availability of reliability as validity test data.

9.2.3   Development of a demographic survey

Part one of the survey should consist of the demographic data collection questions. This part of the survey is as critical as the actual survey used to measure culture and trust. The demographic data collection part of the instrument should be written in such a way as to capture key information such as the following:

- Organization, department, or group name
- Work group name
- Individual job title
- Individual's level within the organization (manager or nonmanager)
- Individual gender
- Individual ethnicity
- Age
- Time working for the organization
- Time in current job
- Job tenure

9.2.4   Developing a survey instrument

Developing a culture or trust survey requires in-depth planning and great thought. The fundamental steps in developing a survey are to define the purpose of the questionnaire, decide what you want to get from it, determine how you will distribute the questionnaire, and the method of analysis. When developing a survey instrument ensure the survey is

- Written in such a way that the respondent is not led to a particular response.
- Contains questions that are clear and easy to understand.
- Contains clear, simple, and consistent responses.
- Completion time 30 minutes or less.
- Survey should be restricted to no more than 50 questions whenever feasible.

9.2.5   Conducting focus group discussions

A focus group is a small group generally consisting of 6 to 12 people brought together to engage in a discussion guided by a trained leader or facilitator. Focus group discussions are used to learn more about the opinions of others on a designated topic. The information gained through these discussions should be used to determine future actions with regard to the topic discussed. The use of focus group surveys and individual interviews is a good combination to collect additional data on the cultural health of an organization. The focus group is a method used to gain insight and understanding about a topic by hearing the perceptions, opinions, or understanding of the topic from a small group of people. The group discussion session should typically be scheduled for one to two hours. Guidelines for conducting focus group surveys are shown below.

- Develop a selection of questions to be asked during the session.
  - Keep the list of questions short, in the range of 10 to 15 questions.
  - Questions should be open-ended conversational type.
  - Questions should be short and clear using terms that are familiar to the participants.
  - Questions should be nonthreatening.
  - Each question should focus on one element or thing.
- Select participants.
  - The group should be kept small, with 6 to 12 participants.
  - Participants should have knowledge of the topic.
- Select the appropriate meeting location.
- Meeting location should be private.

9.2.6   Selecting a facilitator

A focus group facilitator can make or break a focus group because he or she is important in facilitating the exchange of information and ideas. The primary role of the facilitator is to guide and encourage group participation and discussion. Guidelines that can be used when selecting a facilitator are listed below.

- Good communication skills
- Skilled in encouraging people to speak
- A good listener
- Able to control and guide the discussion if necessary
- Skilled in asking scripted predetermined questions and unscripted follow-on questions

- Able to put the participants at ease
- Has experience in facilitating group discussions
- Knows something about the topic
- Able to deal tactfully with outspoken group members

9.2.7   Conducting interviews

Interviews should be used when gleaning information from one individual. This method is the best to use to get feedback from the management team. The same questions discussed during the focus group should be discussed during the interview process. Individual interviews should be scheduled for one hour as a general rule. During the process allow the interviewee ample time to expound on questions and provide complete feedback of his or her observations and perceptions.

9.2.8   Analyzing and reporting the data

Analysis and interpretation of the data should be conducted with transparency to minimize any potential perceived bias. It is also important to understand that the self-assessment was not conducted to be used as a pass/fail mechanism, but as a method of identifying strengths and weaknesses of the organization. Therefore, the results should not be reported as pass/fail. Data analysis can be complex or relatively simple depending on the scope of the assessment, the design of the survey, and the information one is expecting to gain from the assessment as a whole. Therefore, the data analysis should be considered in the initial planning stage of the assessment. Once the data have been collected through means of the survey, focus groups, and the interview process, the data should be analyzed and reviewed to determine areas of strengths and weaknesses. The results reported to management and the employees should occur in a timely fashion.

9.2.9   Culture improvement plan

Once the data have been analyzed and areas for improvement have been identified, it is time to create the culture improvement plan. The improvement plan should address areas that were determined to be below the desired outcome as well as address ways to maintain or improve the areas that were identified as strengths. The improvement plan should provide a detailed roadmap on how the program, process, and behaviors that are in need of improvement should be addressed. Also included in the plan are measures that will be taken to avoid eroding the areas that were identified as strengths.

9.3   Case study 1: Dysfunctional management

Alex, the manager of special projects, works for a national technology firm and is having a complex year meeting commitments for his team and organization. As a result, he enlisted the assistance of his program manager Pauline to help in getting one of his high-profile projects back on track. The project Alex is planning on transferring to Pauline had received a lot of attention from his senior management team because of the project's importance to the success of the company. One of the reasons the project is significantly behind schedule is that Alex has not completely bought into the project and its value to the organization.

During the transfer of the project to Pauline, Alex provided very little information on the many outstanding issues, strategies, and resources needed to get the project back on track. However, he did communicate to her that the project was behind schedule, complaints on the lack of progress were being voiced by the customer, and that he expected that she would resolve all issues and get the project back on track expeditiously. Alex attached no value to the project and believed that the project could not be salvaged therefore he did not want to be directly associated with its failure. The project was in a state of failure because Alex did not provide the attention and resources needed early on or act on issues associated with the project when they occurred.

Pauline, amazed with the information she had received from her manager, began analyzing the project and the associated issues and devised a strategy for success. She realized that if she had a chance of being successful with her new assignment she needed additional resources to assist her. She began to identify and select the appropriate resources she needed to turn the project around. Pauline hired two highly experienced professionals with a great deal of experience and knowledge of the technical aspects of the project.

Months passed and her manager still was uninterested in the project's progress or whether Pauline needed any assistance from him. Pauline continued working the strategy and within approximately six months the project was thriving, receiving awards and accolades from many including the customers. Immediately Alex got on board and began taking what credit he could for the success of the project. He then told Pauline, "I knew you could do it."

After the project was completed Pauline received positive feedback on her performance as a leader from many including the senior executives in her company and her external customers. Her performance impressed Alex's boss to the point that he offered her an opportunity of higher authority on the team. The position that Alex's boss offered Pauline was the position that she had expressed an interest in during many conversations with Alex. Pauline was excited about the opportunity and accepted the position. Alex had worked hard over the years to ensure that Pauline continued to work directly for him. It was well known that Pauline essentially ran the organization and Alex received credit for her work. Therefore, he had a vested interest in keeping her nearby. On one occasion, Alex told Pauline that she was his "best-kept secret." Alex believed that as long as he could keep her under the radar the risk of his losing her through promotion or reassignment was low. Although Pauline continued to work in the same organization with Alex as her manager, she excelled in her new leadership role, meeting and oftentimes exceeding the expectations and needs of her customers.

Case Study 1 Discussion Questions

DISCUSSION QUESTIONS TO CONSIDER

1. How should Alex have handled his discussion with Pauline?
2. How did Alex most likely view the manager/employee relationship?
3. Describe Alex's boss leadership skills.
4. What good leadership traits were displayed by Alex in handling the reassignment of Pauline?
5. What good leadership skills did Pauline display?
6. Describe the characteristics exhibited by Alex that makes him a trusted manager.
7. What do you supposed Alex meant when he referred to Pauline as his "best-kept secret"? How do you think this comment may have affected Pauline (career, perceptions etc.)?

9.4 Case study 2: The absent manager

Donna is a dedicated employee who is always willing to assist with projects and business transactions. She is well known for being a self-starter, an excellent project manager, and strategist. Donna is viewed by many as a good leader who has the ability to inspire the people she leads. She has held essentially every technical and leadership position in her current department. Donna has had numerous discussions with her manager concerning new opportunities in another department. She believed that for the most part, these discussions had fallen on deaf ears because she had not received feedback from her manager. Being persistent, she thought that she would have another discussion with her manager about moving into another position in a different department. The discussion with her manager did not go as well as Donna had expected. Patricia, her manager was agitated and noncommittal during the discussion. After the discussion, Patricia initiated discussions with the human resources manager and others communicating that Donna had not been meeting her expectations in managing her group effectively. Patricia had not previously provided feedback to Donna that she was not meeting expectations and previously rated Donna high on her annual performance review. In addition, Patricia convinced her management that Donna was critical to the organization and should not be permitted to leave the department at the time of the request.

One month later Patricia reassigned Donna to a position of lesser authority. Approximately three months in her new role, Donna was presented with yet another opportunity to move into a position of greater responsibility and authority in another department. Because Donna was no longer in a critical role, Patricia could not justify denying her the opportunity to take advantage of the new position. Therefore, Donna was permitted to move into the new position. As you can imagine, she was excited and quickly accepted the new challenge. Donna reported to her new manager approximately two weeks later and spent approximate three years as the manager of the new group. Her new management was extremely happy with her performance as a leader and her ability to manage resources and achieve results. She was rated high by her new manager on performance reviews as she had been by all of her previous managers including Patricia.

Approximately three years later Patricia posted a management position that Donna thought would be a good leadership opportunity. Therefore, Donna submitted her resume to the human resources department expressing her interest in the position. During the initial screening process, she was informed that her resume would not be forwarded to the hiring manager as she did not meet the minimum requirement for the position. Donna requested that the screening committee take another look at her credentials. After re-evaluating her resume, the screening committee agreed with Donna that she was qualified for the position and that her resume would be forwarded to Patricia, the hiring manager for consideration. Two weeks later, Donna was scheduled for an interview. Days before the interview, Donna was informed by Jim, her current manager, that Patricia (Donna's previous manager) requested that he sit on the interview panel for the position in which she was interested. After receiving the make-up of the interview panel she became perplexed and reluctant to go through with the interview. Although reluctant to participate in the interview process, Donna proceeded to prepare for and was interviewed by the panel consisting of her current boss (Jim) and her previous boss, along with two other panel members. Donna was contacted by Patricia a day later letting her know that she was not selected for the position.

Case Study 2 Discussion Questions

DISCUSSION QUESTIONS TO CONSIDER

1. What type of leader is Patricia?
2. What type of leader is Donna?
3. Did Patricia make any mistakes as a manager? If yes, list them
4. How should Patricia have handled Donna's new opportunity?
5. What would you recommend Patricia do to improve her leadership skills?
6. Describe the relationship between Patricia and Donna.
7. Describe the appropriateness and fairness of the interview process.
8. List the actions taken by management that can serve as a hindrance to trust.
9. Discuss Patricia and Jim in terms of their ability to gain and maintain trust with Donna and the people they lead.

9.5   Case study 3: The art and importance of relationships

Samantha has just accepted a high-profile position in another part of the country. She is apprehensive yet excited and is looking forward to starting her new position. After the announcement was made that she would be joining the staff she began corresponding with some of the key people in her new organization. She gathered and studied key information such as policies, procedures, and practices; organizational structure; and known outstanding issues. Upon arrival at her new location, she immediately met with her senior leadership team. During that discussion she talked briefly about her professional background, provided general information on her immediate family, and discussed her expectations for the organization. Samantha also scheduled and held introductory meetings with her customers internal and external. These meetings were well received and provided a platform for Samantha to begin building a relationship with her customers.

Samantha found her new job to be challenging yet rewarding. She quickly began realizing the importance of relationships in influencing decisions when dealing with both internal and external customers. Realizing that she needed to build a trusting relationship with her new staff she began scheduling team-building sessions with the entire senior leadership team. These sessions were well received by her new staff. Samantha also instituted one-on-one meetings with each of her direct reports to ensure that she kept abreast of progress and issues as well as to provide them the support they needed to be successful. She also met frequently with her manager to keep her informed and to get feedback.

Case Study 3 Discussion Questions

DISCUSSION QUESTIONS TO CONSIDER

1. Discuss Samantha's leadership style and skills.
2. What activities did Samantha engage to get to know her new staff and the customer?
3. What actions did Samantha take to build trusting relationships with her new team?
4. What was Samantha's purpose for discussing her personal background and family with her new staff?

9.6   Case study 4: Diversity matters

Paul and Cassandra are colleagues working for a top engineering firm. Cassandra has an impressive resume with a wide array of experience in the field of engineering. She also has a PhD in systems engineering. Paul has a bachelor's degree in system engineering with some experience in system design. The management team for the engineering firm consists of twenty males and one female. The male:female ratio for the workforce is approximately 80% males and 20% females. The company posted a management position with the intention of filling the job within 30 days from a pool of internal candidates. Both Paul and Cassandra applied for the position and believed that each of them had a shot at successfully landing the position. Cassandra's current boss told her prior to the interview that he did not believe that she would perform well on the interview and that she probably would not fit in with the current leadership team.

The next week Cassandra went through the interview process and felt comfortable that she performed well during the interview. Three days after the interview Cassandra was contacted by the hiring manager and was informed that although the team thought she performed extremely well during the interview process she would not be offered the position. Later that day she discovered that the position was offered to Paul. After accepting the position Paul approached Samantha and told her that he was surprised that he was offered the position because everyone knows that she is better qualified and has consistently demonstrated superior performance since joining the firm.

Case Study 4 Discussion Questions

DISCUSSION QUESTIONS TO CONSIDER

1. What thoughts are going through Cassandra's mind concerning the fairness of the interview process?
2. What message is being sent to Cassandra by her boss about her chances of being successful and being selected for the management position?
3. Can the management team be viewed as trustworthy?
4. Is Cassandra working in a culture that facilitates trust? Explain your answer.
5. What actions displayed by the management team can lead to mistrust?
6. What are Cassandra's perceptions of her organization and the leadership team?

9.7   Case study 5: Management accountability

The manager of nuclear engineering, Paul works for a national technology firm with a reputation of delivering quality services to all of its customers. He leads a group of engineers known to be high performers and who have the reputation for getting the job done on schedule. Paul is known as a micro manager who is intimately involved in every project. At times, his involvement can slow down the delivery of services to the customer. Paul's leadership style is often frustrating to his team members and has led to team members withholding critical project-related information from him.

Paul's manager met with him and counseled him about his inability to meet critical commitments and schedules. Paul immediately proceeded to blame his team for the team's inability to meet scheduled commitments. After the meeting with his manager, Paul scheduled a meeting with his team. During the meeting Paul communicated to his team that he expected them to resolve all of the issues associated with the failing project within the next 30 days otherwise they should begin looking for new jobs. Everyone on Paul's team was aware that the team's inability to meet commitments was due to Paul's extensive hands-on management style. The team worked aggressively over the next 30 days to meet commitments.

Case Study 5 Discussion Questions

DISCUSSION QUESTIONS TO CONSIDER

1. What leadership skills did Paul display?
2. Discuss Paul's competency as a leader.

3. Explain the impact of Paul's leadership style.
4. How did Paul's leadership style affect productivity and communication within the team?

9.8   Individual Trust Questionnaire

The Individual Trust Questionnaire (ITQ) is designed to rate an individual's performance in the area of trust qualitatively. The questionnaire also provides limited measurement of the individual trust attributes (openness and honesty, concern for employees, relationship, competence, reliability, and identification).

9.8.1   Survey completion

Read each question and rate each question on a scale of 1 to 5 by placing an X or a checkmark in the selection box next to the question. The corresponding scales for the questions are listed in Table 9.1.

9.8.2   Survey analysis

Once the survey has been completed it can be analyzed using the following steps.

1. Calculate the result for each scoring category. For example, if answer 3 is selected for 5 questions then the averaging for that category would be 15 (3 × 5).
2. Once all of the categories have been totaled divide the total by the number of questions in the survey (20) to determine the score for the survey. To determine the score for the attributes, divide by the number of attribute questions.
3. The overall score is then compared to Table 9.2 to obtain the trust level for the entire survey.

The above process can be repeated to determine the trust score for the trust attributes using the Individual Trust Questionnaire Guide in Section 9.8.3.

Table 9.1 ITQ Scale

Score	Correspondence results
1	Always
2	Mostly
3	Sometimes
4	Rarely
5	Never

Table 9.2 ITQ Trust Scale

Score	Correspondence results
1	Very High
2	High
3	Moderate
4	Low
5	Very Low

9.8.3 Individual Trust Questionnaire guide

1. I willingly share information without hesitation (openness and honesty).
2. I ensure that the workers are kept informed (concern for employees).
3. I am receptive to new ideas and thoughts that come from others (openness and honesty).
4. I communicate information completely and in totality (openness and honesty).
5. I demonstrate strong leadership and management skills during my daily interactions (competence).
6. I lead with confidence and assurance (competence).
7. I am clear and concise when communicating (openness and honesty).
8. I communicate openly and honestly (openness and honesty).
9. I ensure appropriate and fair compensation for the workers (concern).
10. I address issues and disagreements in a timely manner (relationship).
11. I am decisive and confident in decision making (competence).
12. I show genuine interest in the work being performed by my team and discuss work progress with them (identification).
13. I support a work–life balance concept for my team (concern for employees).
14. I listen actively and openly during communication (relationship).
15. I work to ensure that workers in my organization are provided a safe and clean workplace (concern).
16. Those who have dealt with me know that I keep the promises I make (reliability).
17. I myself follow the policies and procedures set by the company (identification).
18. I demonstrate buy-in for the goal of our organization through both verbal and nonverbal communication and actions (identification).
19. I admit mistakes without fear of being viewed as ineffective (relationship, openness, and honesty).
20. I am consistent and predictable in my actions (reliability).

9.8.4   *Individual Trust Questionnaire*

Individual Trust Questionnaire								
For each question given below, circle the number that best describes your opinion to the questions listed.								
1	**2**	**3**	**4**			**5**		
Always	**Mostly**	**Sometimes**	**Rarely**			**Never**		
#	**Question**			**1**	**2**	**3**	**4**	**5**
1	I willingly share information without hesitation							
2	I ensure that workers are kept informed							
3	I am receptive to new ideas and thoughts that comes from others							
4	I communicate information completely and in totality							
5	I demonstrate strong leadership and management skills during my daily interactions							
6	I lead with confidence and assurance							
7	I am clear and concise when communicating							
8	I communicate completely and honestly							
9	I ensure appropriate and fair compensation for workers							
10	I address issues and disagreements in a timely manner							
11	I am decisive and confident in decision making							
12	I show genuine interest in the work being performed by my team and discuss the work progress with them							
13	I support a work life balance for my team							
14	I listen actively and openly during communication							
15	I work to ensure that the workers in my organization are provided a safe and clean workplace							
16	Those who have dealt with me knows that I keep the promises I make							
17	I myself follow the policies and procedures set by our company							
18	I demonstrate buy-in for the goals of our organization through both verbal and non-verbal communication and actions							
19	I admit mistakes without fear of being viewed as ineffective							
20	I am consistent and predictable in my actions							
Overall Score: _____ Attribute Scores: Openness and Honesty _____ Competence _____ Identification _____ Reliability _____ Relationship _____ Concern for Employees _____								

9.9 Organization Trust Questionnaire

The Organization Trust Questionnaire (OTQ) is designed to rate perception of an organization's performance in the area of trust qualitatively. The questionnaire also provides some measurement of the individual trust attributes (openness and honesty, concern for employees, relationship, competence, reliability, and identification).

9.9.1 Survey completion instructions

Read and rate each question on a scale of 1 to 5 by placing an X or a checkmark in the selection box next to the question. The corresponding scales for the questions are listed in Table 9.3.

9.9.2 Survey analysis

Once the survey has been completed the survey should be analyzed using the following steps:

1. Calculate the result for each scoring category.
2. Once all of the categories have been totaled divide the total by the number of questions to determine the mean score for the survey. To determine the score for the attributes divide by the number of attribute questions.
3. The overall score should be compared to Table 9.4 to obtain the trust level for the entire survey.

Table 9.3 OTQ Scale

Score	Correspondence results
1	Always
2	Mostly
3	Sometimes
4	Rarely
5	Never

Table 9.4 OTQ Trust Scale

Score	Correspondence results
1	Very High
2	High
3	Moderate
4	Low
5	Very Low

The above process can be repeated to determine the trust score for the trust dimension using the Organization Trust Questionnaire Guide.

9.9.3   Organization Trust Questionnaire guide

1. Senior management in my organization communicates information completely and frequently (openness and honesty).
2. Managers in my company support a work–life balance for all workers (concern for employees).
3. The managers in my organization are not afraid to admit when they are wrong (openness and honesty, relationship).
4. Clear and concise communication permeates the organization (openness and honesty).
5. The managers in my organization keep the promises they make (reliability).
6. Disagreements and issues are addressed in a timely manner (relationship).
7. The leaders in my organization are decisive in their decision making (competence).
8. Management provides clear, concise, and honest feedback to subordinates and colleagues (openness and honesty).
9. The managers in my organization demonstrate good leadership and management skills while conducting business and making decisions (competent).
10. Information is communicated in totality without holding back critical elements that are important for employees to know (openness and honesty).
11. My workplace is safe and clean (concern for employees).
12. People in my organization treat each other with respect (relationship).
13. Organization members are willing to share information (openness and honesty).
14. My company benefits program is adequate and comparable to similar companies (concern for employees).
15. The managers in my organization lead with confidence (competence).
16. The managers in the organization listen attentively to workers (relationship).
17. Organization members are receptive to new ideas (openness and honesty).
18. Communication flows in all directions in my organization to ensure that workers are kept informed (concern for employees).
19. People in my organization always follow policies and procedures (identification).

20. The policies and practices in my organization ensure that workers are compensated fairly (concern for employees).
21. My organization is able to retain the skilled workers needed to complete work.

9.9.4 *Organization Trust Questionnaire*

Organization Trust Questionnaire				
For each question given below, circle the number that best describes your opinion to the questions listed.				
1	2	3	4	5
Always	Mostly	Sometimes	Rarely	Never

#	Question	1	2	3	4	5
1	Senior management in my company communicates information completely and frequently					
2	Managers in my company support a work–life balance for all workers					
3	The managers in my organization are not afraid to admit when they are wrong					
4	Clear and concise communication permeates the organization					
5	The managers in my organization keep the promises they make					
6	Disagreements and issues are addressed in a timely manner					
7	The leaders in my organization are decisive in their decision making					
8	Management provides clear, concise, and honest feedback to subordinates and colleagues					
9	The managers in my organization demonstrate good leadership and management skills while conducting business and making decisions					
10	Information is communicated in totality, without holding back critical elements that are important for employees to know					
11	My workplace is safe and clean					
12	People in my organization treat each other with respect					
13	Organization members are willing to share information					
14	My company benefit program is adequate and comparable to similar companies					
15	The managers in my organization lead with confidence					
16	The managers in my organization listens attentively to workers					
17	Organizational members are receptive to new ideas					

Organization Trust Questionnaire - Continued								
For each question given below, circle the number that best describes your opinion to the questions listed.								
1		**2**	**3**		**4**		**5**	
Always		**Mostly**	**Sometimes**		**Rarely**		**Never**	

#	Question	1	2	3	4	5
18	Communication flows in all directions in my organization to ensure that workers are kept informed					
19	People in my organization follow policies and procedures					
20	The policies and practices in my organization ensure that workers are compensated fairly					
21	My organization is able to retain the skilled workers needed to complete work					

Overall Score: _____
Attribute Scores:
Openness and Honesty _____   Competence _____   Identification _____
Reliability _____   Relationship _____   Concern for Employees _____

9.10   Employee Engagement Questionnaire

The Employee Engagement Questionnaire (EEQ) is designed to rate an individual's perception of their engagement in organizational activities qualitatively. The results gained through completion of the questionnaire can provide an indication of whether employees believe they are active participants in the organization.

9.10.1   Survey completion

Read each question and rate each question on a scale of 1 to 5 by placing an X or checkmark in the selection box next to the question. The corresponding scales for the questions are listed in Table 9.5.

9.10.2   Survey analysis

Once the survey has been completed the survey can be analyzed using the following steps:

1. Calculate the result for each scoring category.
2. Once all of the categories have been totaled divide the total by the number of questions to obtain the score for the survey.
3. The overall score should be compared to Table 9.6 to obtain the trust level for the entire survey.

9.10.3 *Employee Engagement Questionnaire*

Employee Engagement Questionnaire				
For each question given below, circle the number that best describes your opinion to the questions listed.				

1	2	3	4	5
Always	Mostly	Sometimes	Rarely	Never

#	Question	1	2	3	4	5
1	I am willing to participate on a team to implement a program or process					
2	My manager often seeks out my opinion					
3	I often have the opportunity to participate in decisions that affect me and the organization					
4	I willingly provide feedback to my managers on issues					
5	I feel valued for the work I perform					
6	I am inspired by my management					
7	I enjoy coming to work					
8	My supervisor provides guidance when needed					
9	I am proud to tell people where I work					
10	Management provides the tools I need to do my job					
11	I understand how my role fits into the organization strategy					
12	Employees are motivated to perform well					
13	Employees in my organization take responsibility for their decisions					
14	My workgroup work together as a team					
15	I have the skills needed to do my job efficiently					
16	I am very satisfied with my job					
17	I would recommend my organization to others as a good place to work					
18	I receive the information I need to perform my job					
19	I share responsibility for the safety of myself and my coworkers					
20	I know what my supervisor expects of me while at work					
21	My coworkers and I are committed to doing a quality job					

Table 9.5 EEQ Scale

Score	Correspondence results
1	Always
2	Mostly
3	Sometimes
4	Rarely
5	Never

Table 9.6 EEG Trust Score

Score	Correspondence results
1	Very High
2	High
3	Moderate
4	Low
5	Very Low

9.11   Management Trust Questionnaire

The Management Trust Questionnaire (MTQ) is designed to gauge qualitatively an individual's perception of trust in management. The results gained from this questionnaire will provide management the knowledge needed to modify actions in order to change workers' perception of the trustworthiness of management.

9.11.1   Survey completion

Read each question and rate each question on a scale of 1 to 5 by placing an X or checkmark in the selection box next to the question. The corresponding scales for the questions are listed in Table 9.7.

9.11.2   Survey analysis

Once the survey has been completed the survey can be analyzed using the following steps:

1. Calculate the result for each scoring category.
2. Once all of the categories have been totaled divide the total by the number of questions to receive the score for the survey.
3. The overall score should be compared to Table 9.8 to obtain the trust level for the entire survey.

9.11.3   Management Trust Questionnaire

Management Trust Questionnaire									
For each question given below, circle the number that best describes your opinion to the questions listed.									
1		**2**		**3**		**4**		**5**	
Always		**Mostly**		**Sometimes**		**Rarely**		**Never**	

#	Question	1	2	3	4	5
1	I feel my manager always tell the truth when communicating to subordinates					
2	My supervisor communicates information openly and honestly					
3	My manager treats me with respect					
4	Management identifies with the goals and value of the organization					
5	Management is consistent and reliable					
6	My manager leads by example and is a good role model for all in the organization					
7	Managers in my organization are viewed as being competent by subordinates					
8	I have a good working relationship with my manager					
9	My manager has the skills to resolve problems to achieve win–win solutions					
10	Managers in my organization operate openly with transparency					
11	There is trust and mutual respect between management and subordinates					
12	My manager consistently works to create an atmosphere of mutual trust					
13	Management does a good job of "walking the talk" on organizational values					
14	My manager treats me fairly					
15	My supervisor is open to hearing and values my opinion					

Table 9.7 MTQ Scale

Score	Correspondence results
1	Always
2	Mostly
3	Sometimes
4	Rarely
5	Never

Table 9.8 MTQ Score

Score	Correspondence results
1	Very High
2	High
3	Moderate
4	Low
5	Very Low

9.12   Instruction for completing the Organizational Culture Questionnaire

The purpose of the Organization Culture Questionnaire (OCQ) is to gauge the type of culture serving as the foundation of the organization's systems. In completing the OCQ you will be providing a picture of important attributes of the organization. The OCQ consists of 28 questions designed to provide an indication of cultural elements applicable to your organization.

9.12.1   Survey instruction

1. Read each question carefully and rank your response based on your perception of organizational performance from 1 to 8.
2. Fill in the response that you believe most closely represents your organization's performance for each question. Please only select one response for each question.
3. Once all questions have been completed use the culture continuum shown in Figure 9.1 to determine where your culture is located based on the average results contained from the response to the questions.

Once the culture mean for the questions has been obtained and compared to the culture continuum, the result can be further clarified and interpreted by using Table 9.9.

9.12.2   Organization Culture Questionnaire

Organization Culture Questionnaire

For each question given below, circle the number that best describes your opinion to the questions listed.

1	2	3	4	5	6	7	8
Always	Mostly	Frequently	Usually	Sometimes	Infrequently	Seldom	Never

#	Question	1	2	3	4	5	6	7	8
1	My organization has a clear vision								
2	The values of my organization are shared by its members								
3	I believe that my management values my opinion								
4	Communication in my organization is fluent and flows in all directions								
5	The managers in my organization recognize and celebrate the success of its members								
6	The mission and values of my organization are posted for employees to view								
7	In my organization management celebrates the successes of employees at every level								
8	The management team is trusted and respected by employees at every level								
9	Management is responsive to suggestions from employees								
10	Conflicts are handled openly and fairly								
11	Employees are motivated to perform their jobs								
12	Employees understand their job duties and their role within the organization								

Organization Culture Questionnaire - Continued

For each question given below, circle the number that best describes your opinion to the questions listed.

1	2	3	4	5	6	7	8
Always	Mostly	Frequently	Usually	Sometimes	Infrequently	Seldom	Never

#	Question	1	2	3	4	5	6	7	8
13	Downward communication is accurate								
14	The organization goals and objectives are clear to employees throughout the organization								
15	Roles and responsibilities within the organization are clear and understood								
16	My input is valued by my peers								
17	Employees have the right training and skills to perform their jobs								
18	Knowledge and information sharing is a common practice for members of my organization								
19	Disagreements are addressed promptly when they occur								
20	Morale is high across my organization								
21	Employees enjoy coming to work								
22	I feel that I am valued as a part of my team								
23	Employees speak highly of my organization								
24	Roles and responsibilities in my organization are clearly defined and understood								
25	Everyone takes responsibility for their actions								
26	My supervisor is a positive role model								

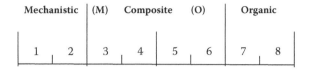

Figure 9.1 Culture continuum scale.

Table 9.9 Culture Type

Score	Culture type	Characteristics
1–2	Mechanistic	Mostly mechanistic characteristics
3–4	Composite	More mechanistic characteristics than organic
5–6	Composite	More organic characteristics than mechanistic
7–8	Organic	Mostly organic characteristics

9.13   Culture focus group questions

1. What do you like most about working for your organization?
2. What three words would you used to describe the culture of your organizations?
3. What are the most common complaints workers have concerning your organization?
4. How does communication flow in your organization?
5. Are you provided the equipment and training needed to perform your job successfully?
6. How comfortable are people in your organization in communicating with management (supervisor, middle management, senior manager)?
7. Are the skills and abilities of the people in my organization valued by management and used in their current job?
8. How importance is trust to the members of your organization?
9. Are employees encouraged and willing to get involved in solving issues?
10. What five words would you use to describe the management team?

9.14   Trust focus group questions

1. How would you describe communication in your organization?
2. What process or practices are in place to encourage open and honest communication between management and subordinates?
3. Describe the behavior of the management team.
4. List 3 trustworthy behavior exhibited by your immediate supervisor or manager.

5. Describe how your manager interacts with your work group.
6. Do you believe that trust is an important attribute for managers during interactions?
7. What is your perception of the actions taken by management to ensure that they are viewed as trust worthy?
8. Describe how employees are treated in the organization.
9. How comfortable are you in discussing issues with management when they arise?
10. Describe how issues are handled when they arise.

9.15 Practitioner guide summary

As we established throughout the book to this point, managers are the backbone of an organization. The actions of leaders are attentively observed and scrutinized every day by colleagues and subordinates. Knowing this, managers must remain cognizant of their actions and the effect of those actions have on others and the success of the organization. Dysfunctional managers produce dysfunctional organizations. In these types of organizations you can expect cultures that are not conducive to building and maintaining trust. One may also expect that these types of organizations have a difficult time meeting commitments and retaining skilled knowledge workers. The Practitioner Guide provides important attributes and characteristics that managers should practice and perfect in their daily lives as leaders. The attributes and characteristics presented will definitely help form the foundation of creating and trusting cultures.

Appendix: Participating organization data and charts

Organization D

Table A.1 Sample Demographics—Organization D

Demographics	% Population
Females	20.0
Males	80.0
African Americans	34.3
Caucasians	51.4
Hispanic	8.6
Other Races	5.7
<25 years of age	47.1
25–35 years of age	32.9
36–45 years of age	15.7
46–55 years of age	4.3
55+ years of age	0.0
<1 year with company	28.6
1 to <5 years with company	52.9
5 to <10 years with company	7.1
10 to <20 years with company	7.1
20 to <30 years with company	2.9
30 or more years with company	0.0
Management	24.3
Nonmanagement	75.7

Table A.2 Trust and Culture Means—Gender
Organization D

Demographics	Trust means	Culture means
Gender		
Females	3.0	3.9
Males	3.3	4.4
Race		
American Indians	3.3	4.5
Caucasians	3.2	4.2
Hispanics	3.3	4.7

Table A.3 Trust and Culture Means—Age Group
Organization D

Age (Years)	Trust means	Culture means
<25	3.2	4.5
25–35	3.2	4.1
36–45	3.2	4.7
46–55	3.9	4.0
55+	NA	NA

Table A.4 Job Tenure Trust and Culture Means—Organization D

Organization tenure	Trust means	Culture means
<1 year	3.2	4.6
1 to <5	3.2	4.3
5 to <10	3.0	3.7
10 to <15	NA	NA
15 to <20	3.7	3.8
20 to <30	3.4	5.0
30 or more	NA	NA

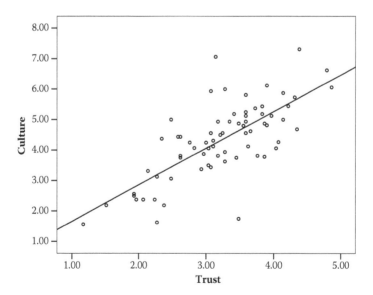

Figure A.1 Scatterplot—Organization D.

Organization E

Table A.5 Sample Demographics E

Demographics	% Population
Females	40.0
Males	60.0
African Americans	0.0
Caucasians	100
<25 years of age	NA
25–35 years of age	NA
36–45 years of age	NA
46–55 years of age	NA
55+ years of age	NA
<1 year with company	NA
1 to <5 years with company	NA
5 to <10 years with company	NA
10 to <20 years with company	NA
20 to <30 years with company	NA
30 or more years with company	NA
Management	20.0
Nonmanagement	80.0

Table A.6 Trust and Culture Means—Gender
Organization E

Demographics	Trust means	Culture means
Gender		
Females	3.6	5.4
Males	3.7	5.6
Race		
Caucasians	3.7	5.6
African Americans	NA	NA

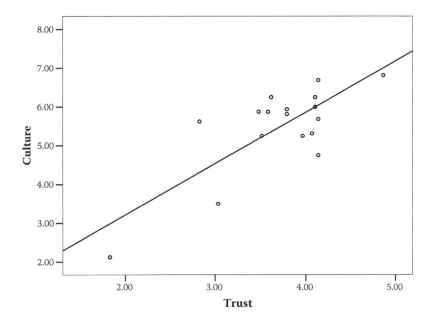

Figure A.2 Scatterplot—Organization E.

Organization F

Table A.7 Sample Demographics—Organization F

Demographics	% Population
Female	26.2
Male	72.7
African American	15.6
Caucasian	78.9
Asian	1.2
American Indian	1.2
Hispanic	0.4
<25 years of age	0.0
25–35 years of age	3.1
36–45 years of age	24.6
46–55 years of age	43.4
55+ years of age	15.6
<1 year with company	0.4
1 to <5 years with company	2.3
5 to <10 years with company	3.1
10 to <20 years with company	44.9
20 to <30 years with company	32.8
30 or more years with company	3.5
Management	17.6
Nonmanagement	78.1

Table A.8 Trust and Culture Means—Gender
Organization F

Demographics	Trust means	Culture means
Gender		
Female	3.7	4.9
Male	3.7	4.9
Race		
African American	3.6	5.0
Caucasian	3.6	4.9
American Indian	3.7	5.0
Asian	4.1	5.5

Table A.9 Trust and Culture Means—Age Group Organization F

Age (Years)	Trust means	Culture means
<25 years	NA	NA
25–35	3.4	4.2
36–45	3.6	4.8
46–55	3.6	4.8
55+	3.7	5.0

Table A.10 Job Tenure Trust and Culture Means—Organization F

Organization tenure	Trust means	Culture means
<1 year	4.4	6.7
1 to <5	3.3	4.5
5 to <10	3.6	4.4
10 to <15	3.7	4.8
15 to <20	3.7	4.8
20 to <30	3.7	5.0
30 or more	3.5	5.2

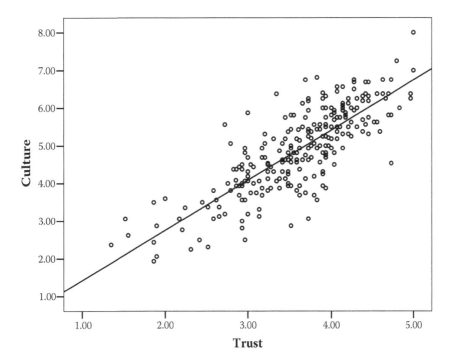

Figure A.3 Scatterplot—Organization F.

Organization G

Table A.11 Sample Demographics—Organization G

Demographics	% Population
Females	54.3
Males	45.7
African Americans	62.9
Caucasians	37.1
<25 years of age	5.7
25–35 years of age	17.1
36–45 years of age	31.4
46–55 years of age	34.3
55+ years of age	8.6
<1 year with company	0.0
1 to <5 years with company	20.0
5 to <10 years with company	40.0
10 to <20 years with company	28.6
20 to <30 years with company	8.6
30 or more years with company	2.9
Management	20.0
Nonmanagement	80.0

Table A.12 Trust and Culture Means—Gender
Organization G

Demographics	Trust means	Culture means
Gender		
Females	2.5	3.2
Males	2.9	3.9
Race		
African Americans	2.5	3.3
Caucasians	3.0	3.9

Table A.13 Trust and Culture Means—Age Group Organization G

Age (Years)	Trust means	Culture means
<25 years	1.2	1.0
25–35	2.4	2.9
36–45	2.6	3.4
46–55	3.0	4.1
55+	3.5	4.5

Table A.14 Job Tenure Trust and Culture Means—Organization G

Organization tenure	Trust means	Culture means
1	NA	NA
1 to <5	2.0	3.2
5 to <10	2.7	3.4
15 to <20	2.9	3.5
20 to <30	3.5	5.1
30 or more	1.8	3.7

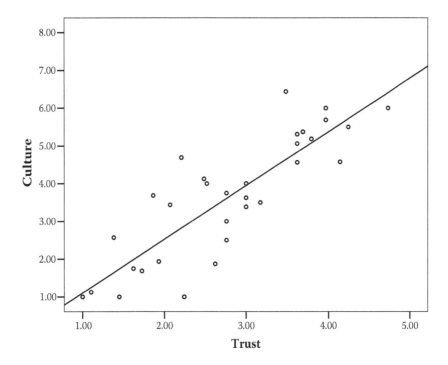

Figure A.4 Scatterplot—Organization G.

Appendix: Participating organization data and charts

Organization H

Table A.15 Sample Demographics—Organization H

Demographics	% Population
Females	92
Males	0.8
African Americans	26.1
Caucasians	73.9
<25 years of age	0.0
25–35 years of age	13.0
36–45 years of age	43.5
46–55 years of age	30.4
55+ years of age	13.0
<1 year with company	17.4
1 to <5 years with company	65.2
5 to <10 years with company	4.3
10 to <20 years with company	4.3
20 to <30 years with company	0.0
30 or more years with company	8.7
Management	8.3
Nonmanagement	91.7

Table A.16 Trust and Culture Means—Gender
Organization H

Demographics	Trust means	Culture means
Gender		
Females	3.7	5.3
Males	4.8	5.2
Race		
African Americans	4.2	5.4
Caucasians	3.6	5.3

Table A.17 Trust and Culture Means—Age Group
Organization H

Age (Years)	Trust means	Culture means
<25 years	NA	NA
25–35	3.8	5.6
36–45	3.8	5.3
46–55	3.7	5.2
55+	4.4	5.1

Table A.18 Job Tenure Trust and Culture Means—Organization H

Organization tenure	Trust means	Culture means
<1 year	4.2	5.9
1 to <5	3.6	5.3
5 to <10	3.7	3.4
10 to <15	3.7	4.9
15 to <20	NA	NA
20 to <30	NA	NA
30 or more	5.0	4.4

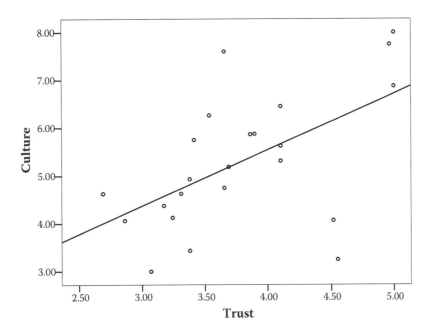

Figure A.5 Scatterplot—Organization H.

Organization I

Table A.19 Sample Demographics—Organization I

Demographics	% Population
Females	33.3
Males	66.7
African Americans	0.0
Caucasians	93.3
Hispanic	6.7
<25 years of age	0.0
25–35 years of age	53.3
36–45 years of age	20.0
46–55 years of age	13.3
55+ years of age	6.7
<1 year with company	6.7
1 to <5 years with company	33.3
5 to <10 years with company	20.0
10 to <20 years with company	33.3
20 to <30 years with company	0.0
30 or more years with company	0.0
Management	73.3
Non-management	26.7

Table A.20 Trust and Culture Means—Gender
Organization I

Demographics	Trust means	Culture means
Gender		
Females	3.8	5.8
Males	4.0	5.4
Race		
African Americans	NA	NA
Caucasians	3.9	5.5
Hispanic	4.0	5.6

Table A.21 Trust and Culture Means by Age
Group—Organization I

Age (Years)	Trust means	Culture means
<25	NA	NA
25–35	3.9	5.5
36–45	3.7	6.0
46–55	3.6	5.2
55+	4.0	5.6

Table A.22 Job Tenure Trust and Culture
Means—Organization I

Organization tenure	Trust means	Culture means
<1 year	3.7	6.1
1 to <5	3.9	5.3
5 to <10	4.2	5.8
10 to <15	3.8	5.5
15 to <20	4.0	5.9
20 to <30	NA	NA
30 or more	NA	NA

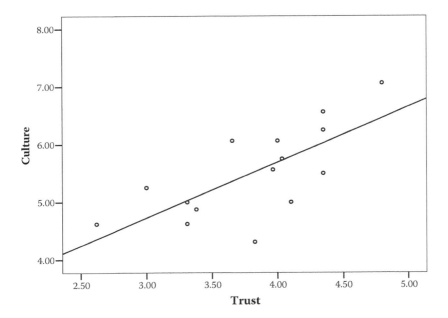

Figure A.6 Scatterplot—Organization I.

Organization J

Table A.23 Sample Demographics—Organization J

Demographics	% Population
Females	14.1
Males	84.7
African Americans	17.6
Caucasians	70.6
Asian	1.2
Hispanic	1.2
<25 years of age	7.1
25–35 years of age	37.6
36–45 years of age	25.9
46–55 years of age	16.5
55+ years of age	4.7
<1 year with company	2.4
1 to <5 years with company	22.4
5 to <10 years with company	35.3
10 to <20 years with company	34.1
20 to <30 years with company	4.7
30 or more years with company	0.0
Management	17.6
Nonmanagement	82.4

Table A.24 Trust and Culture Means—Gender
Organization J

Demographics	Trust means	Culture means
Gender		
Females	3.0	4.0
Males	3.3	4.2
Race		
African Americans	3.1	4.2
Caucasians	3.3	4.2
Asian	3.5	4.1
Hispanic	3.7	4.4

Table A.25 Trust and Culture Means—Age Group Organization J

Age (Years)	Trust means	Culture means
<25	3.0	4.5
25–35	3.4	4.3
36–45	3.3	4.2
46–55	3.4	4.0
55+	4.0	4.1

Table A.26 Job Tenure Trust and Culture Means—Organization J

Organization tenure	Trust means	Culture means
<1 year	3.9	4.8
1 to <5	3.0	4.4
5 to <10	3.4	4.1
10 to <15	3.2	3.7
15 to <20	3.5	4.3
20 to <30	3.1	4.0
30 or more	NA	NA

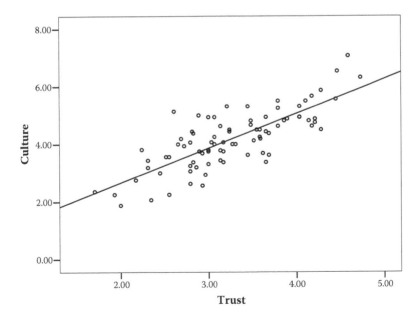

Figure A.7 Scatterplot—Organization J.

Organization K

Table A.27 Sample Demographics—Organization K

Demographics	% Population
Females	13.3
Males	86.7
African Americans	33.3
Caucasians	66.7
<25 years of age	0.0
25–35 years of age	0.0
36–45 years of age	26.7
46–55 years of age	46.7
55+ years of age	26.7
<1 year with company	0.0
1 to <5 years with company	0.0
5 to <10 years with company	13.3
10 to <20 years with company	13.3
20 to <30 years with company	60.0
30 or more years with company	13.3
Management	6.7
Nonmanagement	93.3

Table A.28 Trust and Culture Means—Gender
Organization K

Demographics	Trust means	Culture means
Gender		
Females	3.1	4.6
Males	4.0	5.3
Race		
African Americans	3.9	5.6
Caucasians	3.7	5.0

Table A.29 Trust and Culture Means—Age
Group Organization K

Age (Years)	Trust means	Culture means
<25 years	NA	NA
25–35	NA	NA
36–45	4.2	5.2
46–55	3.6	5.7
55+	3.6	4.3

Table A.30 Job Tenure Trust and Culture
Means—Organization K

Organization tenure	Trust means	Culture means
<1 year	NA	NA
1 to <5	NA	NA
5 to <10	4.4	4.8
10 to <15	NA	NA
15 to <20	4.1	5.5
20 to <30	3.5	5.4
30 or more	4.2	4.1

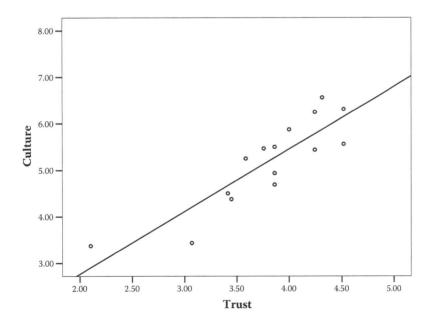

Figure A.8 Scatterplot—Organization K.

Organization L

Table A.31 Sample Demographics—Organization L

Demographics	% Population
Females	45.5
Males	54.5
African Americans	9.1
Caucasians	90.9
<25 years of age	9.1
25–35 years of age	9.1
36–45 years of age	36.4
46–55 years of age	27.3
55+ years of age	18.2
<1 year with company	9.1
1 to <5 years with company	36.4
5 to <10 years with company	9.1
10 to <20 years with company	9.1
20 to <30 years with company	36.4
30 or more years with company	0.0
Management	9.1
Nonmanagement	90.9

Table A.32 Trust and Culture Means—Gender Organization L

Demographics	Trust means	Culture means
Gender		
Females	3.4	5.4
Males	3.8	4.8
Race		
African Americans	3.8	6.7
Caucasians	3.6	4.9

Table A.33 Trust and Culture Means—Age Group Organization L

Age (Years)	Trust means	Culture means
<25 years	3.8	6.7
25–35	4.0	5.5
36–45	3.2	4.4
46–55	3.9	5.4
55+	3.8	4.9

Table A.34 Job Tenure Trust and Culture Means—Organization L

Organization tenure	Trust means	Culture means
<1 year	3.8	6.7
1 to <5	3.5	5.1
5 to <10	3.6	5.5
10 to <15	3.9	4.3
15 to <20	NA	NA
20 to <30	3.5	4.7
30 or more	NA	NA

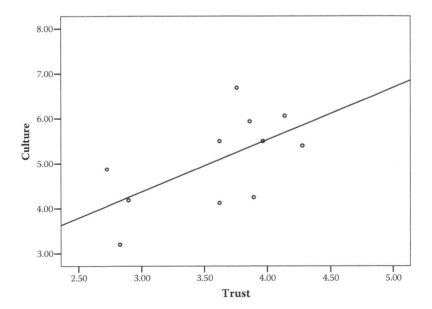

Figure A.9 Scatterplot—Organization L.

Organization M

Table A.35 Sample Demographics—Organization M

Demographics	% Population
Females	50
Males	50
African Americans	8.8
Caucasians	91.2
<25 years of age	0
25–35 years of age	23.5
36–45 years of age	35.3
46–55 years of age	29.4
55+ years of age	8.8
<1 year with company	0
1 to <5 years with company	17.6
5 to <10 years with company	17.6
10 to <20 years with company	50
20 to <30 years with company	11.8
30 or more years with company	2.9
Management	14.7
Nonmanagement	85.3

Table A.36 Trust and Culture Means—Gender Organization M

Demographics	Trust means	Culture means
Gender		
Females	3.7	5.0
Males	3.7	5.4
Race		
African Americans	4.5	5.3
Caucasians	3.6	5.2

Table A.37 Trust and Culture Means by Age
Group—Organization M

Age (Years)	Trust means	Culture means
<25 years	NA	NA
25–35	3.7	5.8
36–45	3.7	5.0
46–55	3.6	5.1
55+	3.5	4.7

Table A.38 Job Tenure Trust and Culture
Means—Organization M

Organization tenure	Trust means	Culture means
<1 year	NA	NA
1 to <5	3.7	5.8
5 to <10	4.0	5.7
10 to <15	4.1	5.0
15 to <20	3.6	5.3
20 to <30	3.5	4.1
30 or more	2.1	2.8

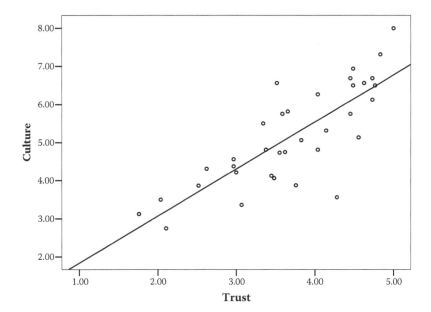

Figure A.10 Scatterplot—Organization M.

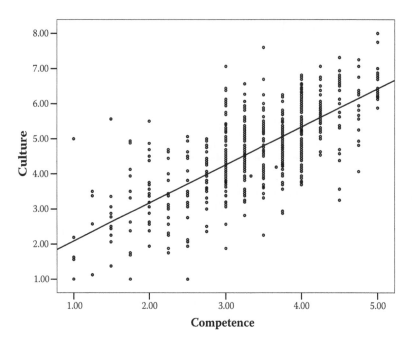

Figure A.11 Scatterplot for competence and culture data.

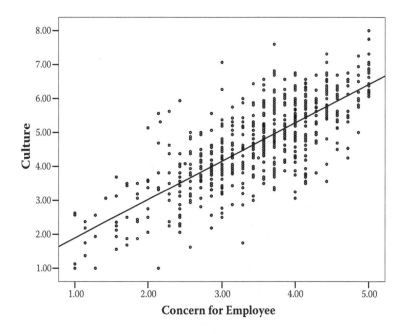

Figure A.12 Scatterplot for concern for employee and culture data.

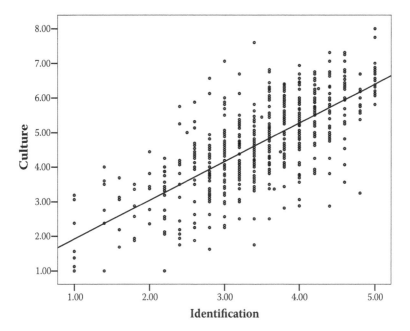

Figure A.13 Scatterplot for identification and culture data.

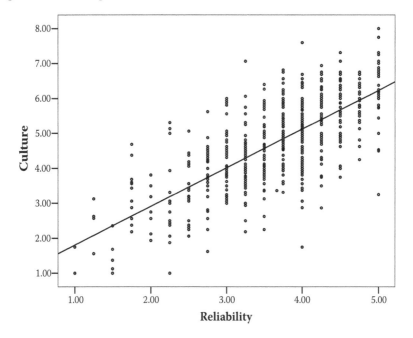

Figure A.14 Scatterplot for reliability and culture data.

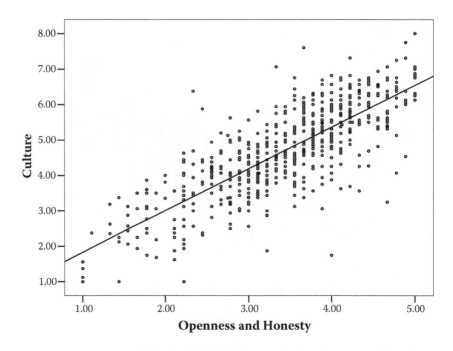

Figure A.15 Scatterplot for openness and honesty and culture.

References

Albrecht, S. L. 2002. Perceptions of integrity, competence, and trust in senior management as determinants of cynicisms toward change. *Public Administration and Management: An Interactive Journal*, 7, 4, 320–343.

Alvesson, M. 1993. *Cultural Perspectives on Organizations*. Cambridge, UK: Cambridge University Press.

Amsa, P. 1986. Organizational culture and workgroup behavior: An empirical study. *Journal of Management Studies*, 23, 3.

Alston, F. 2007. The relationship between perceived culture and trust in technology-driven organizations, dissertation. Huntsville, AL: University of Alabama.

Basier, A. 1986. Trust and antitrust. *Ethics*, 96, 2.

Bender, R., Neuhauser, P. C., and Stromberg, K. L. 2000. *Culture.Com*. Hoboken, NJ: John Wiley & Sons.

Bennis, W. 2000. *Managing the Dream Reflections on Leadership and Change*. New York: Perseus Publishing.

Blanchard, K., and Hodges, P. 2003. *The Servant Leader*. Nashville, TN: Thomas Nelson Inc.

Blonqvist, K. 1997. The many faces of trust. *Scandinavian Journal of Management*, 13, 3, 271–286.

Burns, T., and Stalker, G. M. 1961. *The Management of Innovation*. London: Tavistock.

Butler, J. K. 1991. Toward understanding and measuring conditions of trust: Evolution of a condition of trust inventory. *Journal of Management*, 17 (3).

Daley, D. M., and Vasu, M. L. 1998. Fostering organizational trust in North Carolina: The pivotal role of administrators and political leaders. *Administration & Society*, 30, 62.

Davis, J., Schoorman, D., Mayer, R., and Tan, H. H. 2000. The trusted general manager and business unit performance: Empirical evidence of a competitive advantage. *Strategic Management Journal*, 25, 5, 563–576.

Deal, T. E., and Kennedy, A. A. 1982. *Corporate Culture, the Rites and Rituals of Corporate Life*. New York: Perseus Publishing.

Denison, D. R. 1990. *Corporate Culture and Organizational Effectiveness*. Hoboken, NJ: John Wiley & Sons.

Dotlich, D. L., and Cairo, P. C. 2002. *Unnatural Leadership Going against Intuition and Experience to Develop Ten New Leadership Instincts*. New York: Jossey-Bass.

Fairholm, G. W. 1994. *Leadership and the Culture of Trust*. Westport, CT: Praeger Publishers.

George, J. M., and Jones, G. R. 1996. *Understanding and Managing Organizational Behavior*. London: Addison-Wesley Publishing Company.

Gilbert, J. A., and Tang, T. L. -P. 1998. An examination of organizational trust antecedents. *Public Personnel Management*, 27, 3, 321.

Gudykunst, W. B., Stewart, L. P., and Toomey, S. 1985. *Communication, Culture, and Organizational Processes*. New York: Sage Publications.

Hampden-Turner, C. 1992. *Creating Corporate Culture*. London: Addison-Wesley Publishing Company Inc.

Harris, P. R. 1989. *High Performance Leadership: Strategies for Maximum Career Productivity*. Glenview, IL: Scott Foresman & Company.

Harrison, R., and Stokes, H. 1992. *Diagnosing Organizational Culture*. New York: Jossey-Bass/Pfeiffer.

Hersey, P. 2004. *The Situational Leader*. Center for Leadership Studies.

Heskett, J., and Kotter, J. P. 1992. *Corporate Culture and Performance*. New York: The Free Press.

Herzberg, F., 1996. *Work and the Nature of Man*. New York: World Publishing Company.

Hofstede, G. 2001. *Culture Consequences*, 2nd ed. Thousand Oaks, CA: Sage Publications.

Hyer, B. 2002. *Building Trust: How to Get It! How to Keep It!* Hyer Bracey, Inc.

Jobber, D., and Horgan, I. G. 1998. A comparison of techniques used and journals taken by marketing researchers in Britain and the USA. *Service Industries Journal*, 8, 3.

Johns, G. 1992. *Organizational Behavior: Understanding Life at Work*, 3rd ed. New York: Harper Collins Publishers.

Kotter, J. P., and Heskett, J. L. 1992. *Corporate Culture and Performance*. New York: The Free Press.

Kouzes, J. M., and Posner, B. Z. 1987. *The Leadership Challenge*. New York: Jossey-Bass Publishers.

Kramer, R. M., and Tyler, T. R. 1996. *Trust in Organizations Frontiers of Theory and Research*. Thousand Oaks, CA: Sage Publications.

Lane, C., and Bachmann, R. 1998. *Trust within and between Organizations*. Oxford: Oxford University Press.

Likert, R. 1967. *The Human Organization: Its Management and Value*. New York: McGraw-Hill Book Company.

Mayer, R. C., and Davis, J. H. 1995. An integrative model of organizational trust. *Academy of Management Review*, 20, 3, 709–734.

Mishra, A. K. 1992. Organizational responses to crisis: The role of mutual trust and top management teams, dissertation. Ann Arbor, MI: School of Business Administration, The University of Michigan.

National Science Foundation. 2012. Science & engineering indicators.

Pearce, J. A., II., and Robinson, R. B., Jr. 1997. *Strategic Management*. New York: McGraw-Hill/Irwin.

Pheysey, D. C. 1993. *Organizational Cultures Types and Transformations*. New York: Routledge.

Reigle, R. F. 2003. Organizational culture assessment: Development of a descriptive test instrument, dissertation. Tuscaloosa, AL: University of Alabama.

Reina, D. S., and Reina, M. L. 1993. *Trust & Betrayal in the Workplace*. San Francisco, CA: Berrett-Koehler Publishers, Inc.

Robbins, S. P. 1984. *Essentials of Organizational Behavior*. Englewood Cliffs, NJ: Prentice Hall Publishing Company.

Robbins, S. P. 1997. *Managing Today*. Englewood Cliffs, NJ: Prentice Hall Publishing Company.

Rotter, J. B. 1967. A new scale for the measurement of interpersonal trust. *Journal of Personality*, 35, 651–665.

Ryan, K. D., and Oestreich, D. K. 1998. *Driving Fear Out of The Workplace: Creating the High-Trust, High-Performing Organization*. New York: Jossey Bass Publishers.

Sackmann, S. A. 1991. *Cultural Knowledge in Organizations Exploring the Collective Mind*. Thousand Oaks, CA: Sage Publications.

Sathe, V. 1983. Implications of corporate culture: A manager's guide to action. *Organizational Dynamics* (fall).

Schein, E. H. 1992. *Organizational Culture and Leadership*, 2nd ed. New York: Jossey-Bass Publishers.

Schein, E. H. 2010. *Organizational Culture and Leadership*. Hoboken, NJ: Jon Wiley & Sons.

Schoderbek, P. P., Cosier, R. A., and Aplin, J. C. 1991. *Management*, 2nd ed. San Diego, CA: Harcourt Brace Jovanovich Publishers.

Shaw, R. B. 1997. *Trust in The Balance Building Successful Organizations on Results, Integrity, and Concern*. New York: Jossey-Bass Publishers.

Shockley-Zalabak, P., Ellis, K., and Cesaria, R. 1999. Measuring organizational trust: Trust and distrust across cultures. The Organizational Trust Index, IABC Research Foundation.

Tang, J. A., and Li-Ping, T. 1998. An examination of organizational trust anteced-ents. *Public Personnel Management*, 27, 3.

Taylor, J. C., and Bowers, D. G. 1972. Survey of organizations: A machine-scored standardized questionnaire instrument. Ann Arbor, MI: Center for Research on Utilization of Scientific Knowledge, University of Michigan.

Tozer, J. 1997. *Leading Initiatives*. Oxford, UK: Butterworth Heinemann.

Trice, H. M., and Beyer, J. M. 1993. *The Cultures of Work Organizations*. Englewood Cliffs, NJ: Prentice Hall Publishing Company.

Tschannen-Moran, M. 2001. Collaboration and the need for trust. *Journal of Educational Administration*, 39, 4, 308–331.

Tway, D. C., Jr. 1994. A construct of trust, dissertation. Austin, TX: The University of Texas at Austin.

Woolston, R. L. 2001. Faculty perceptions of dean transitions: Does trust matter? An interpretive case study of organizational trust and organizational culture, dissertation. San Diego, CA: School of Education, University of San Diego.

Zammuto, R. F., and Krakower, J. Y. 1991. Quantitative and qualitative studies of organizational culture. *Research in Organizational Change and Development*, 5, 83–114.

Zand, D. E. 1997. *The Leadership Triad*. Oxford, UK: Oxford University Press.

Index